TRINIDAD AND TOBAGO MATHS CONNECT 1

Irmin Edwards-Kennedy

Yvonne Fitz-Andrews-Ramsey

Gwendolyn Pope

Series editor:
Karen Morrison

HODDER EDUCATION
AN HACHETTE UK COMPANY

Although every effort has been made to ensure that website addresses are correct at time of going to press, Hodder Education cannot be held responsible for the content of any website mentioned in this book. It is sometimes possible to find a relocated web page by typing in the address of the home page for a website in the URL window of your browser.

Hachette UK's policy is to use papers that are natural, renewable and recyclable products and made from wood grown in well-managed forests and other controlled sources. The logging and manufacturing processes are expected to conform to the environmental regulations of the country of origin.

Orders: please contact Bookpoint Ltd, 130 Park Drive, Milton Park, Abingdon, Oxon OX14 4SE. Telephone: +44 (0)1235 827827. Fax: +44 (0)1235 4004501. Email education@bookpoint.co.uk Lines are open from 9 a.m. to 5 p.m., Monday to Saturday, with a 24-hour message answering service. You can also order through our website: www.hoddereducation.com

First published by Pearson Education Limited
Published from 2015 by Hodder Education,
An Hachette UK Company
Carmelite House
50 Victoria Embankment
London EC4Y 0DZ
www.hoddereducation.com

20

10 9 8 7 6 5 4 3 2

British Library Cataloguing in Publication Data is available
from the British Library on request.

ISBN: 978 0 435891 53 4

Typeset by Tech-Set Ltd, Gateshead, Tyne and Wear
Original illustrations © Harcourt Education Limited, 2006
Illustrated by Adrian Barclay and Tech-Set Ltd.
Cover design by mccdesign ltd.
Printed in Dubai

Cover photo: © Getty images

Acknowledgements
Every effort has been made to contact copyright holders of material reproduced in this book. Any omissions will be rectified in subsequent printings if notice is given to the publishers.

Maths Connect for Trinidad and Tobago has been designed to support the Secondary Education Modernization Programme (SEMP) for the lower secondary school maths curriculum of Trinidad and Tobago. It has built on the research and consultation with maths teachers, educators and curriculum officers across the Caribbean region that was conducted for *Caribbean Maths Connect* and in the UK for *Maths Connect*. The publishers would like to acknowledge the contribution made by the authors of *Caribbean Maths Connect*: Benita Byer-Bowen, Rosalyn Kelly and Delia D. Samuel.

The publishers would like to thank the following for permission to use photographs:
p1 Harcourt Education Ltd/Mike van der Wolk, p5 The Seattle Times/Ken Lambert, p22 Harcourt Education Ltd/Mica Brancic, p34, p41, p42, p43 Harcourt Education/Mike van der Wolk, p44 Harcourt Education Ltd/Mica Brancic, p55 Harcourt Education Ltd/Mike van der Wolk, p91 Corbis, p110, p113, p121 Harcourt Education Ltd/Mike van der Wolk, p152 AKG, p161 Alamy Images

The authors and publishers would like to thank Dirk Strasser, Geoff Phillips Publications Trust and Nolan Creative Services Pty Lts for permission to use examples from Chapter Review Sections from **Heinemann Outcomes Mathematics 7 CSF II Edition**

and SSMG for permission to use projects on pages 76, 89 and 138 from **Heinemann Mathematics 9 Extension textbook**.

Contents

To the teacher

Trinidad and Tobago Maths Connect is a unique course organised to fit the **General Curriculum Framework for Trinidad and Tobago (Form 1)**. It covers key areas of knowledge, understanding and skills which provide a firm foundation to raise competence and confidence in mathematics. The lively approach of the text is designed to make the study of mathematics both interesting and enjoyable.

The table on the right indicates how the material in this course fits with the General Curriculum Framework for Trinidad and Tobago (Form 1).

Trinidad and Tobago Maths Connect, with its wealth of stimulus material, will help students to understand the relevance of mathematics to everyday problems. There are sample investigations, and suggestions for project work which follow each chapter. The book also contains illustrated case studies to show how mathematics is used by real people in different jobs.

The modules are designed to be used sequentially. Each 'Looking back' relates to the previous section, and 'Revise and consolidate' applies to all previous material. The 'Test your knowledge' tests at the end relate to topics throughout the book.

Students should be able to use calculators where appropriate, but they should not become over-dependent on them.

How to use this book

This book is divided up into 16 colour-coded **chapters**. Number chapters are yellow; Algebra chapters are green; Geometry chapters are blue; Measurement chapters are purple and Data Handling chapters are orange. Each chapter is divided up into **units**.

Unit 1.2 Understanding number systems

In this unit you will:

- find out more about early number systems
- read and write numbers using early number systems
- describe the decimal or denary system and how it works
- understand the place value of digits in whole numbers
- carry out research to find out more about our number system.

Key words
Digits
Numeral
Place value
Denary
Decimal

Directly under the unit heading are the **learning outcomes**. These tell you what you will learn in the lesson. To help you remember important vocabulary, there is a **Key words** box here.

> A number is an amount. The symbol used to represent a number is called a **numeral**. In our number system we have ten numerals or **digits**: 1, 2, 3, 4, 5, 6, 7, 8, 9, 0.

In the **explanation box**, you can see a summary of the key ideas that are covered in the lesson. The key words are printed in bold type.

Worked examples

1. What is the value of the underlined digit in each number?

a. <u>2</u> **b.** <u>9</u>00 **c.** <u>8</u> 000 **d.** <u>6</u>0

2. Write down a six-digit number and give the value of the third and fourth digit.

1. **a.** Two **b.** Nine hundred **c.** Eight thousand **d.** Sixty

2. 34<u>5</u> <u>9</u>87

5 – five thousands = 5 000

9 – nine hundreds = 900

The **worked examples** show you how to answer the exercise questions. On the blue paper you can see the kind of working you should be writing in your exercise book. The green **hint boxes** help explain how to calculate the answers.

The **exercises** are made up of three types of questions:

- **practice** questions, which allow you to practise the basic skills
- **problem** questions, which ask you to apply the skills you have learned
- **investigation** questions, which give you practice at solving open-ended problems

At the end of each chapter are 'Looking back' exercises to help you review what you have learned. There are also seven 'Revise and consolidate' exercises spread throughout the book for more review and practice work, and finally, ten 'Tests of your knowledge' at the end.

Key words
Numbers

Numbers and number systems

In this chapter you will learn about early number systems, the ways in which these used place value, and how to read and write numbers using different number systems. This will help you to understand some of the characteristics of the denary number system that we use today as well as to read, write and classify different types of numbers.

Unit 1.1 Numbers all around us

In this unit you will:

- read and write numbers.

Why are numbers so important to us?

If you look around you it's very likely that you will see at least one number in use. In the modern world, we all use **numbers** every day without really thinking about them. In fact, Brian Butterworth in his book *The Mathematical Brain* says: "At a very, very rough guess, I would say that I process about 1 000 numbers per hour, about 16 000 numbers per waking day, nearly 6 million a year. People whose job entails working with numbers in supermarkets, banks, betting shops, schools and dealing rooms, will process many more than this." The pictures show you a few of the places in which numbers are used in our daily lives.

Exercise

1. Say each number shown in the pictures in words.
2. Write down ten more examples of how we use numbers on a daily basis.

Unit 1.2 Understanding number systems

In this unit you will:

- describe the decimal or denary system and how it works
- understand the place value of digits in whole numbers
- carry out research to find out more about our number system.
- find out more about early number systems
- read and write numbers using early number systems

Key words
Numeral
Number system
Digits
Denary
Decimal
Place value

A number is an amount. The symbol used to represent a number is called a **numeral**. In our **number system** we have ten numerals or **digits**: 1, 2, 3, 4, 5, 6, 7, 8, 9, 0. These are arranged next to each other to form larger numbers.
567 represents 5 hundreds, 6 tens and 7 units.
2 099 represents 2 thousands, 0 hundreds, 9 tens and 9 units (or ones).

Our **denary** number system uses a base 10 or decimal counting system. Denary means divided or counted by tens. **Decimal** is another word for denary. The base 10 system probably developed because people have ten fingers and ten toes which they can use to help them count.

In the denary or decimal system the position, or place, of each digit determines its value. **Place value** increases in multiples of 10 from right to left.

Hundred thousands $10 \times 10 \times 10 \times 10 \times 10$ or 10^5	Ten thousands $10 \times 10 \times 10 \times 10$ or 10^4	Thousands $10 \times 10 \times 10$ or 10^3	Hundreds 10×10 or 10^2	Tens 10×1 or 10^1	Units 1 or 10^0
2	2	<u>2</u>	2	2	2

Each of the 2s in this number has a different place value. The place value of the number tells you the value of the number. The underlined two has a place value of thousands and an actual value of <u>2</u> 000. Its face value is 2.

Zero is used as a place holder.

Worked examples

1. What is the value of the underlined digit in each number?
   a. <u>2</u> **b.** <u>9</u>00 **c.** <u>8</u> 000 **d.** <u>6</u>0

2. Write down a six-digit number and give the value of the third and fourth digit.

1. **a.** Two **b.** Nine hundred **c.** Eight thousand **d.** Sixty

2. 34<u>5</u> <u>9</u>87
   5 – five thousands = 5 000
   9 – nine hundreds = 900

Exercise

1 Write the value of each underlined digit in words and numerals.

a) $\underline{2}03$
b) $4\underline{6}\,89\underline{1}$
c) $\underline{1}0\,005$
d) $\underline{2}92\,9\underline{1}1$
e) $\underline{8}4$
f) $6\,\underline{1}51$
g) $\underline{1}\,502\,\underline{8}00$
h) $\underline{7}4\underline{9}\,212$

2 The table shows the number of tourists arriving on five islands during one year.

Island	Tourist arrivals
St John	429 300
Bear Island	311 781
River Island	54 821
St Peterstown	305 167
Jamestown	98 099

a) Which island had the largest number of tourist arrivals?

b) Which island received the fewest tourists?

c) Which island received three hundred and five thousand, one hundred and sixty-seven tourist arrivals?

3 The population of some Caribbean countries (in 2004) is shown in the table:

Country	Population
Jamaica	2 700 000
Trinidad and Tobago	1 310 000
Barbados	268 000
St Lucia	158 000
Belize	247 000

a) Write the populations in order from smallest to largest.

b) Write the population of each country in words.

Investigation

1 Carry out your own research to find out more about the history of our number system. If you have access to the Internet you can use your search engine to find some starting points.

2 What is the metric system? How is it similar to the denary system we use for counting?

3 Our number system is a base 10 system. The ancient Sumerians used a base 60 system. Work in pairs to design a base 6 number system. Write down three four-digit numbers using your base 6 system. Swap these with another group and work out the value of each other's numbers.

Unit 1.3 Different types of numbers

In this unit you will:

- identify, use and classify different types of numbers.

Key words
Whole number
Natural number
Even Odd
Prime Composite
Rectangular number

Natural or counting numbers start at 1 and continue to infinity.
N = {1, 2, 3 …}

Whole numbers start at 0 and continue to infinity. W = {0, 1, 2, 3 …}

Even numbers are divisible by 2. For example 4 and 718.

Odd numbers are not exactly divisible by 2. For example 3 and 449.

Prime numbers can be divided by only two numbers – the number itself and 1. For example 3, 11 and 23.

Any number that is not prime is a **rectangular number**. This means we can write it as a product of two factors and represent it as a rectangle. For example 12 is a rectangular number which can be written as 1 × 12, 2 × 6 or 3 × 4. We can draw dots to show these as rectangles:

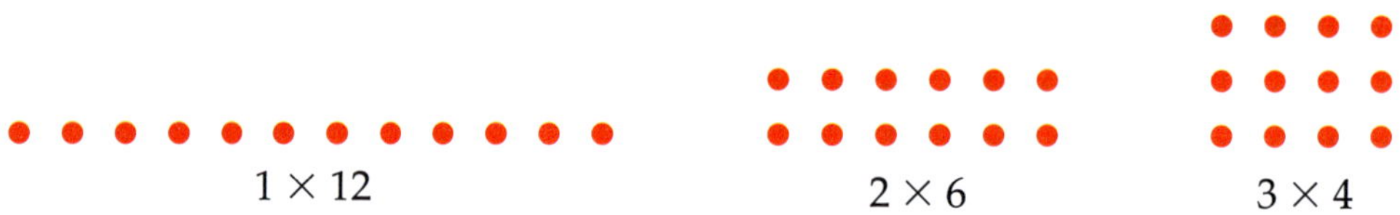

Composite numbers can be divided by more than two numbers. For example 6 can be divided by 1, 2, 3 and 6; 25 can be divided by 1, 5 and 25.

Exercise

1. Copy and complete the table by ticking the correct column(s).

Number	Whole	Natural	Odd	Even	Prime	Rectangular	Composite
8							
17							
45							
257							
21							
29							
$\frac{7}{8}$							
$4\frac{1}{5}$							
64							
1							

2. Is 1 a prime number? Give a reason for your answer.

3. What does infinity mean? Is infinity a number? Is infinity a place?

4. Are all even numbers composite? Give a reason for your answer.

5 Write the following sets of numbers.

a) Prime numbers less than 20.

b) Composite numbers between 7 and 29.

c) Odd numbers between −20 and −40.

d) Rectangular numbers between 17 and 39.

6 Say whether the answer will be an odd or even number in the following operations.

a) odd + odd b) odd × odd c) even + even

d) even × even e) odd × even f) odd + even

Investigation

The Greek scholar Eratosthenes, who lived around 200 BC, developed a method to identify prime numbers less than a certain number *n*. His method is called the sieve of Eratosthenes. It works like this:

- List the numbers from 2 to 100. (You can use this system up to any value.)
- Circle the smallest prime number (2) and then cross out all the numbers divisible by it.
- Circle the next prime number (3) and then cross out all the numbers divisible by it.
- Continue like this till you have circled or crossed out all the whole numbers on your list. The prime numbers are the ones you have circled.

Today mathematicians use computers to find larger and larger primes. In June 2004, NewScientist.com announced that Josh Findley of Seattle, USA had found the largest known prime number. It had 7 million digits! This is a million digits more than the last large prime discovered!

Findley used his home desktop computer, which was linked to a special project called the Great Internet Mersenne Prime Search (GIMPS). This project has signed up more than 200 000 personal computers. Volunteers download software that runs in the background and uses spare processing power to test whether a number is prime.

You can find out more about the latest on prime numbers by entering 'largest prime number' into your Internet search engine. In February 2005, Dr. Martin Nowak of Germany found the new largest prime number, which has 816 230 digits more than Josh Findley's. It is the 42nd Mersenne prime.

1 Use the sieve of Eratosthenes to find all the prime numbers less than 100.

2 Carry out your own research into Eratosthenes. Use it to write a paragraph which outlines his other mathematical achievements.

3 Find out where the name 'Mersenne prime number' comes from and why these primes are so special.

4 The Goldbach conjecture is a statement that suggests that all even numbers greater than 2 can be written as the sum of two prime numbers. This has never been proven mathematically. Why do you think it's so difficult to prove this?

Unit 1.4 Writing numbers in index form

In this unit you will:

- Convert numbers to index form
- Convert indices to a whole number.

Key words
Number
Base
Index
Squared
Power
Cubed

We can write **numbers** in different ways. For example we can write 3×3 as 3^2.

The raised 2 is index notation. It tells us how many times to multiply 3 by itself. In this example, 3 is the **base** and 2 is the **index**.

We read 3^2 as 'three **squared**' or 'three to the **power** of two'.

We can use any number as the index. For example:

$3^1 = 3$

$3^3 = 3 \times 3 \times 3$ — This is read as 'three **cubed**' or 'three to the power of three'.

$3^6 = 3 \times 3 \times 3 \times 3 \times 3 \times 3$ — This is read as 'three to the power of six'.

Worked examples

1. Write the following using index notation:
   a. $4 \times 4 \times 4$ b. $5 \times 5 \times 5 \times 5$ c. 100
2. Express 1 000 using powers of 10.
3. Express 16 using powers of 2.
4. Find the value of the following:
   a. 2^3 b. 4^5 c. 10^6

1. a. 4^3 b. 5^4 c. 100^1
2. $1\,000 = 10 \times 10 \times 10 = 10^3$
3. $16 = 2 \times 2 \times 2 \times 2 = 2^4$
4. a. $2^3 = 2 \times 2 \times 2 = 8$ b. $4^5 = 4 \times 4 \times 4 \times 4 \times 4 = 1\,024$
   c. $10^6 = 10 \times 10 \times 10 \times 10 \times 10 \times 10 = 1\,000\,000$

Exercise

1. Write the value of:

   a) 2^2 b) 3^2 c) 3^3

   d) 7 squared e) 5 cubed f) eight to the power of two

   g) ten to the power of three h) 3^5 i) 10^9

2. Write the following as powers of ten:

   a) one hundred b) ten thousand c) one million

   d) one hundred thousand e) one billion f) ten

   g) one

3. Copy and complete the table showing the first ten square numbers:

Index form	**Expanded form**	**Square numbers**
1^2	1×1	1
2^2		
3^2		
4^2		
5^2		
	6×6	
	7×7	
	8×8	
		81
		100

4. Draw up your own table to show the first ten cube numbers.

5. Who am I?

   a) Double me is four to the power of two.

   b) One tenth of me is ten to the power of two.

   c) I am the difference between two cubed and two squared.

   d) I am five to the power of two added to ten to the power of three.

Investigation

Some numbers are equal to the sum of two square numbers.

For example, $34 = 3^2 + 5^2$ and $13 = 2^2 + 3^2$

Which numbers less than 100 can be expressed as a sum of two square numbers?

Unit 1.5 The metric system

Key words
Units
Metre

In this unit you will:

- **Make a connection between the metric system and the place value system**
- **Read and write the relationship between sub-units of the metric system.**

You have seen that the denary system works on the base ten and that we can use place value to create larger and larger numbers. For example, we can represent the denary system up to seven places in a table like this one:

1 000 000	100 000	10 000	1 000	100	10	1
10^6	10^5	10^4	10^3	10^2	10^1	10^0

The denary system also forms the basis for the metric system that we use for measurement. If we place the **units** of length into the table above, we can read off the relationship between different sub-units of the metric system:

1 000 000	100 000	10 000	1 000	100	10	1
10^6	10^5	10^4	10^3	10^2	10^1	10^0
km	hm	dam	m	dm	cm	mm

Converting between metric units of length is relatively easy because we only ever need to divide or multiply by powers of ten to change to different units. When you are converting to a smaller unit, you need to multiply. When you are converting to a larger unit, you need to divide.

Worked example

Convert:

a. 62 km to m **b.** 256.5 m to mm **c.** 875 cm to km

Look at the table. You need to move three places to the right to get from kilometres to metres. This means you have to multiply by 10^3 or 1 000.

a. 62 km × 1 000 = 62 000 m

Look at the table. You need to move three places to the right to get from metres to millimetres. This means that you have to multiply by 10^3 or 1 000.

b. 256.5 m × 1 000 = 256 500 mm

Look at the table. You need to move five places to the left to get from centimetres to kilometres. This means that you need to divide by 10^5 or 100 000.

c. 875 cm ÷ 100 000 = 0.00875 km

Investigation

Copy the place value table above and complete it for mass and volume.

Work in pairs.

Discuss the relationships between the sub-units of the metric system and write down as many relationships between the units as you can

Exercise

1. Copy and complete the following conversions.

a) 5 km = _______ m
b) 23 km = _______ m
c) 3.6 km = _______ m
d) 9.7 km = _______ m
e) 0.2 km = _______ m
f) 0.003 km = _______ m
g) 700 m = _______ cm
h) 9 m = _______ cm
i) 3 m = _______ cm
j) 0.55 m = _______ cm
k) 5 t = _______ kg
l) 647 t = _______ kg
m) 8 kg = _______ g
n) 0.033 kg = _______ g

1 tonne = 1 000 kg

2. Copy and complete.

a) 4.7 km = _______ cm
b) 9.05 km = _______ cm
c) 0.7 km = _______ mm
d) 0.342 km = _______ mm
e) 4.95 m = _______ mm
f) 3.2 m = _______ mm
g) 7.1 t = _______ g
h) 0.35 t = _______ g

3. Copy and complete the following conversions.

a) 67.6 mm = _______ cm
b) 9.17 kg = _______ t
c) 455 g = _______ kg
d) 800 cm = _______ m
e) 5 000 kg = _______ t
f) 7.25 m = _______ km
g) 6 000 m = _______ km
h) 4 000 mm = _______ m
i) 9 000 mm = _______ m
j) 500 mm = _______ m
k) 33 mm = _______ m
l) 7 000 g = _______ t

4. A tree in the rainforest in Tobago was measured to be 13 258 cm high. How many metres is this?

5. The bird-eating spider of South America has a body length of 89 mm and a leg span of 254 mm. Write these measurements in centimetres.

6. The mass of the average male human brain is 1 408 g. What is this in kilograms?

Looking back

1 Write the following numbers in figures using the denary system.

a) two thousand three hundred and seventy

b) ten thousand and twenty-seven

c) nine million three hundred and seventy-two thousand

d) one hundred and seventy-three thousand three hundred and thirty-three

2 Write down the actual value of the 6 in each number.

a) 4 678 **b)** 1 007 768 **c)** 1 678 002 **d)** 9 089 226

3 What is the difference between the face value, place value and actual value of a digit?

4 Look at the numbers in the box.

a) List the composite numbers.

b) Write down a number that is prime and even.

c) Which of these numbers are prime?

1	2	9	17	23	54
61	63	83	99	107	127

5 How do mathematicians today find large prime numbers?

6 Express the following in index form.

a) $2 \times 2 \times 2 \times 2$

b) $4 \times 4 \times 4$

c) $10 \times 10 \times 10 \times 10 \times 10$

d) $12 \times 12 \times 12 \times 12 \times 12 \times 12$

e) five times five

f) ten times ten times ten

7 What is the value of:

a) 2^1 **b)** 2^3 **c)** 2^{10}

d) 10^4 **e)** 10^0 **f)** 10^1

g) 5^3 **h)** 3^3 **i)** 9^3?

8 Copy and complete the following table.

Kilometres	Metres	Centimetres	Millimetres
12			
99			
100			
2 200			

Chapter 2

Key words
Addition
Subtraction
Multiplication
Division
Total
Sum
Difference

Working with numbers

This chapter deals mainly with using the four operations to solve problems. You will use the order of operations and recognise and apply various mathematical laws to estimate, solve and check answers. Then you will build number sequences and work with number patterns. You will also learn more about factors and multiples and use your knowledge to solve related problems.

Unit 2.1 Operations on numbers

In this unit you will:

- perform addition, subtraction, multiplication and division to solve problems
- use a calculator to check your results.

To add a set of whole numbers use the column method. Place the numbers in columns, one below the other, making sure that the digits are aligned by place value. Add from the right, regrouping to the column to the left as necessary.

```
   15 972
      459
+ 136 801
---------
  153 232
```

Other words that tell us to use **addition** are **sum** and **total**. To subtract whole numbers, write the numbers in the column method and then subtract from the right. Regroup from the column to the left as necessary.

```
  34 901
- 28 472
--------
   6 429
```

Another word that tells us to use **subtraction** is **difference**.

Worked example

Mrs James has $578 dollars in her bank account at the start of June.

a. On the 5th she deposits $200 dollars into her account. On the 7th she deposits another $143. What is her new balance?

b. On the 22nd Mrs James withdraws $354 to pay for an air ticket. How much money is left in her account?

a. 578 + 200 + 143 = □

```
  578
  200
+ 143
-----
  921
```

Her new balance is $921.

b. 921 − 354 = □

```
  921
- 354
-----
  567
```

She has $567 left in her account.

Exercise

1. Work out the following additions.

   a) 15 052 + 4 894 + 36 701 + 87
   b) 789 252 + 476 189 + 34 900
   c) 99 999 + 88 888 + 77 777 + 66 666
   d) 4 073 + 14 670 + 265 + 140
   e) 25 341 + 36 127 + 48 659

2. Perform the following subtractions.

   a) 15 052 − 14 670
   b) 4 073 − 1 999
   c) 211 001 − 187 925
   d) 10 000 − 3 912
   e) 89 322 − 69 887

3. Use a calculator to check your answers for questions 1 and 2.

4. The table shows the number of visitors arriving at St David on various airlines in one year.

   a) Which airline carried the most passengers to St David?
   b) Which airline carried the lowest number of arrivals?
   c) Find the total number of arrivals at St David for this year.
   d) What is the difference between the numbers of arrivals using:
      i) North Air and South Air?
      ii) North Air and Skyway?

Airline	Arrivals
South Air	152 491
North Air	87 849
Fly Anytime	27 430
Up and Away	153 102
Skyway	68 565

5. a) Find the difference in price between the most expensive house and the least expensive house.
   b) Mr Joseph is a property developer. He bought White Walls and Pearl House. What was the total cost of both houses?
   c) Mr Joseph does renovations to both houses and sells them each for $25 500 more than he paid for them. What are the new selling prices of the houses?

White Walls

Red Mansion

Yellow Bricks

Pearl House

6. The table shows the most populated islands in the Caribbean as estimated in 2003.

   a) Write each population out in full in figures.
   b) What is the total population of these six islands?
   c) What is the difference in the populations of:
      i) Cuba and Trinidad and Tobago?
      ii) Puerto Rico and Jamaica?

Island	Population in thousands
Haiti	8 100
Dominican Republic	8 500
Trinidad and Tobago	1 310
Puerto Rico	3 950
Jamaica	2 700
Cuba	11 222

Multiplication and division

For simple **multiplication** and **division** you should know your tables and perform the operation mentally.

Today we have calculators to help us with more difficult calculations, but we still need to be able to do multiplications and divisions using pen and paper methods.

```
    4 512
×      39
---------
  135 360
   40 608
---------
  175 968
```

```
         1 569      ← quotient
     23)36 089      ← dividend
divisor →23
         13 0
         11 5
          1 58
          1 38
           209
           207
             2      ← remainder
```

For a quick method of multiplying by 10, 100 or 1 000, the digits move one, two or three places to the left on a place value grid. Zero is used as a place holder.

When dividing by 10, 100 or 1 000, the digits move one, two or three places to the right on a place value grid.

Th	H	T	U	
		2	3	
	2	3	0	23 × 10
2	3	0	0	23 × 100
	2	3	0	2 300 ÷ 10
		2	3	230 ÷ 10

Worked example

A secretary can type 50 words per minute.

a. How many words can she type in 45 minutes at the same rate?

b. How long will it take her to type 1 800 words?

a. 50 x 45 = □

```
   50
 x 45
 ----
2 250
```

She can type 2 250 words in 45 minutes.

b. 1 800 ÷ 50 = □

```
      36
 50)1 800
```

It will take her 36 minutes.

Exercise

1. Perform these multiplications.
   - a) 1 572 × 38
   - b) 637 × 24
   - c) 1 895 × 45
   - d) 312 × 14
   - e) 129 × 109
   - f) 1 234 × 100

2. Work out the following divisions. Identify the dividend, divisor, quotient and remainder.
   - a) 3 465 ÷ 15
   - b) 835 ÷ 26
   - c) 4 000 ÷ 32
   - d) 11 430 ÷ 18
   - e) 6 031 ÷ 48
   - f) 900 ÷ 100

3. Check your answers to question 2 using a calculator.

4. Using the numbers 99 and 125, find:
   - a) the total
   - b) the difference
   - c) the product
   - d) the quotient

5. Find the missing values.
   - a) □ × 12 = 10 740
   - b) □ ÷ 16 = 145
   - c) 6 240 ÷ □ = 240
   - d) □ × 22 = 1 562
   - e) What does this reveal about division and multiplication?

6. A taxi driver uses 7 litres of petrol for every 100 km that he travels. How many litres would be need for a trip of:
   - a) 100 km
   - b) 1 000 km
   - c) 550 km?

7. An aeroplane carries baggage with a maximum mass of 1 980 kilograms. For a flight to Puerto Rico there are 13 suitcases with a mass of 32 kilograms each, 18 suitcases that each have a mass of 29 kilograms and 34 suitcases each with mass 27 kilograms.

   - a) What is the total mass of the suitcases on this flight?
   - b) The airline agrees to carry another 13 boxes of equal mass to get to the maximum allowed baggage. What should the mass of each of these boxes be?

8. Sheila is training for the Commonwealth Games. Each morning she swims 4 800 metres in a pool that is 25 metres long.
   - a) How many lengths of the pool does she swim?
   - b) How far will she have swum in total in 10 days?
   - c) How far will she swim if she swims the same distance every day for the month of August?

Investigation

1. Find 148 792 × 25 and check your answer on a calculator.

   There is a quicker method for multiplying whole numbers by 25. We know that 25 = 100 ÷ 4, so to find 148 792 × 25, first multiply 1 478 792 by 100 then divide the result by 4. Try it! Now do these using the quick method:
   - a) 3 754 × 25
   - b) 876 512 × 25. Check your answers on your calculator.

2. Use a quick method to find **a)** 467 × 125 **b)** 5 486 × 125. Check your answers on your calculator.

Key words
Operation
Commutative
Associative
Distributive

Unit 2.2 Order of operations

In this unit you will:

- identify which operations you should do first (×, ÷, +, −)
- use the commutative, associative and distributive laws.

When there is more than one **operation** in a calculation, we have to perform each calculation in the correct order.

The expression within brackets must be calculated first. The answer should be inserted in the same position in the expression.

When you find the word 'of' in an expression, it means multiply. For example, $\frac{1}{2}$ of 4 means $\frac{1}{2} \times 4$.

When an expression includes addition and subtraction and multiplication and division, you should work from left to right. Do the divisions and multiplications first, then the additions and subtractions.

$$6 + 4 \times 7 - 13$$
$$\downarrow$$
$$6 + 28 - 13$$
$$\downarrow$$
$$34 - 13$$
$$\downarrow$$
$$21$$

Addition and multiplication are **commutative**. This means it does not matter which order you do them in: $2 + 3 = 3 + 2$ and $2 \times 3 = 3 \times 2$

Subtraction and division are not commutative. The order in which you do them is important. $4 - 2 \neq 2 - 4$ and $4 \div 2 \neq 2 \div 4$

Addition and multiplication are also **associative** operations. It does not matter how numbers are grouped: $5 + (10 + 3) = (5 + 10) + 3$ and $(4 \times 7) \times 2 = 4 \times (7 \times 2)$

Subtraction and division are not associative.

Multiplication is **distributive** over addition and subtraction. This means that in an expression like $7 \times (3 + 5)$, both the 3 and the 5 must be multiplied by 7.
In other words: $7 \times (3 + 5) = (7 \times 3) + (7 \times 5)$

Worked examples

1. $3 \times (13 - 8) + 4$

$3 \times (13 - 8) + 4$
$= 3 \times 5 + 4$
$= 15 + 4$
$= 19$

2. $64 + (25 - 13) \div 6 - 8 \times 5$

$64 + (25 - 13) \div 6 - 8 \times 5$
$= 64 + 12 \div 6 - 8 \times 5$
$= 64 + 12 \div 6 - 40$
$= 64 + 2 - 40$
$= 66 - 40$
$= 26$

Exercise

1. Work out the answer to these calculations. Rewrite each calculation in a different order and then check that both sets of answers are the same.

   a) $8 + 7 + 3$ b) $4 + 5 + 9 + 6$ c) $47 + 21 + 33 + 19$
   d) $17 - 5 - 6 - 1$ e) $12 - 3 - 4 - 5$ f) $19 - 1 - 6 - 2$
   g) $14 - 5 + 2$ h) $11 + 2 - 8$

2. Work out the answer to these calculations. Rewrite each calculation in a different order and then check that both sets of answers are the same.

   a) $5 \times 2 \times 3$ b) $6 \times 4 \times 2$ c) $7 \times 3 \times 2$
   d) $8 \times 5 \times 3$ e) $16 \div 2 \div 4$ f) $48 \div 3 \div 4$
   g) $160 \div 5 \div 10$ h) $144 \div 4 \div 12$

3. Calculate:

   a) $4 \times 8 - 1$ b) $7 + 6 \times 8 - 3$ c) $12 - 8 \div 2 + 3$
   d) $3 \times 9 + 8 \times 6$ e) $4 \times 2 - 3 \div 1 + 6$ f) $14 \times 8 - 1 + 7$

4. These are the answers to the calculations in the questions below:

18	19	40	27	9	4	19	6
25	7	8	20	15	44	5	

   However, an extra answer has been included by mistake. Complete each calculation and find the extra answer.

   a) $6 \times 4 + 3$ b) $3 \times 5 + 10$ c) $16 - 5 \times 2$ d) $4 \times 2 - 1$
   e) $2 \times 8 + 3$ f) $12 + 3 \times 2$ g) $11 - 6 \times 1$ h) $4 \times 2 + 7$
   i) $5 \times 4 - 1$ j) $24 - 2 \times 10$ k) $37 + 35 \div 5$ l) $29 - 27 \div 3$
   m) $8 + 4 \times 0$ n) $10 + 6 \times 5$

5. Write 'true' or 'false' for each of these. Do the calculations and check whether you were correct. Remember to apply the rules you have learned.

   a) $12 - 7 + 6 = 12 + 6 - 7$ b) $7 \times 12 \div 6 = 12 \div 6 \times 7$
   c) $9 + 8 - 5 - 2 = 9 - 8 + 5 + 2$ d) $12 \div 6 + 30 = 12 + 30 \div 6$
   e) $11 - 7 \times 3 = 11 \times 3 - 7$ f) $35 \div 5 \div 2 = 35 \div 2 \div 5$
   g) $24 \div 2 \times 4 = 24 \times 4 \div 2$ h) $16 + 4 \times 3 - 2 = 16 + 3 \times 4 - 2$

Investigation

Calculate $3 \times 7 + 4 \times 8$.

Investigate different ways of writing this calculation without changing the value of the answer. For example: $8 \times 4 + 7 \times 3$.

Choose your own calculation with mixed operations, and investigate different ways of writing the same calculation.

Key words
Estimate
Approximate
Round

Unit 2.3 Estimation

In this unit you will:

- estimate by rounding numbers to the nearest power of 10
- approximate and estimate to check reasonableness of answers.

An **estimate** is a 'guess' or **approximate** value. We make estimations all the time. For example, we estimate how much people weigh or how old they are. The population figures you worked with in Chapter 1 are also estimates because it's almost impossible to give an exact value to the number of people in a country.

In maths you will **round** numbers to make your estimates as accurate as possible. To round a number you have to follow some maths rules:

- Look at the digit in the place you wish to round to.
- If the digit to the right of this digit is 4 or less, it will not change.
- If the digit to the right of this digit is 5 or greater, add 1 to the digit you are rounding to.
- **Replace all digits to the right with zeros**.

361 to the nearest 10 $\approx$ 360 (the digit in the units column is < 4)
368 to the nearest 10 $\approx$ 370 (the digit in the unit column is $\geqslant 5$, so we add 1)
2 050 to the nearest 100 $\approx$ 2 100 (the digit in the tens column is $\geqslant 5$, so we add 1)

The symbol $\approx$ means 'almost equal to' or 'approximately'.

Worked examples

1. Round 38 254 to the:

a. nearest hundred **b.** nearest thousand

a. 38 254
The next digit is 5, so we add 1.
38 300

b. 38 254
The next digit is 2 so it stays the same.
38 000

2. Estimate and add 3 568, 809 and 3 814.

3 600 + 800 + 3 800 = 8 200

$$\begin{array}{r} 3568 \\ 809 \\ +\,3814 \\ \hline 8191 \end{array}$$

The answer is close to the estimate so it's probably correct.

Exercise

1 Round the prices of the cars to:

a) the nearest ten **b)** the nearest hundred **c)** the nearest thousand

2 Estimate the answers to the following by rounding each number to the nearest ten. Calculate the actual answers and compare them with your estimates.

a) 36 + 48 **b)** 52 − 39 **c)** 121 × 13

d) 27 × 68 **e)** 1 288 ÷ 23 **f)** 735 ÷ 21

g) Find the difference between your estimate and the actual answer in each case.

3 Give the approximate distance from:

a) Roxborough to Charlotteville

b) Scarborough to Roxborough

c) Charlotteville to Crown Point

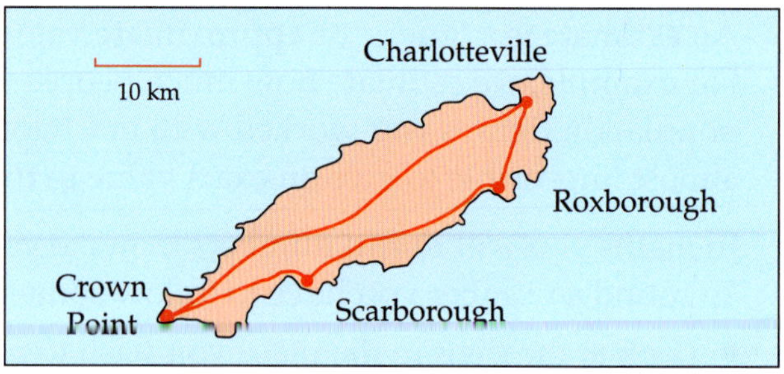

4 Choose the best estimate from those given. Do not do the actual calculations.

a) 324 × 74 — 210 643 210 2 163 21 000

b) 6 500 × 77 — 480 000 420 000 350 000 42 000 560 000

c) 7 564 ÷ 83 — 100 1 000 640 000 560 000 10

d) 126 905 ÷ 9 499 — 100 10 10 000 300 500

5 When people work with a calculator it's easy to make mistakes. Explain how estimating the answer can help you use the calculator more efficiently.

Investigation

1 Work in groups of six. You will need a tape measure and a stop watch. Record all your results.

a) Estimate the height (in metres) of each person in your group.

b) Use the tape measure to find each person's actual height.

c) Estimate (in seconds) how long it will take each person in your group to run around the school building.

d) Use the stop watch to find the exact times.

e) Compare your estimates with your measurements. How good were your estimates? How could you improve them?

2 Estimate how many lines there are in this diagram. How can you work out how many lines there are without counting them?

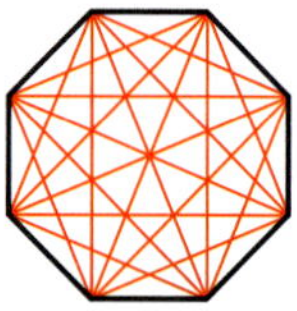

Key words
Sequence
Term
Ascending
Descending

Unit 2.4 Sequences and number patterns

In this unit you will:

- recognise the rules used to generate various sequences and number patterns
- complete sequences and number patterns and generate your own.

A number **sequence** is an ordered set of numbers which follow a rule or pattern.

Each number in a sequence is called a **term**.

Sequences may be in **ascending** order (going up), for example 2, 4, 6, 8 …, or **descending** order (going down), for example 18, 15, 12, 9 …

Worked examples

Find the next two terms in these sequences.

a. 4, 9, 14, 19, 24, ☐, ☐ **b.** 15, 12, 9, 6, ☐, ☐ **c.** 1, 3, 9, 27, ☐, ☐

a. There is a difference of 5 between the terms. The sequence is ascending so we need to add 5.

$24 + 5 = 29$

$29 + 5 = 34$

The next two terms are 29 and 34.

b. There is a difference of 3 between the terms. The sequence is descending so we need to subtract 3.

$6 - 3 = 3$

$3 - 3 = 0$

The next two terms are 3 and 0.

c. To get from one term to the next, multiply by 3.

$27 \times 3 = 81$

$81 \times 3 = 243$

So the missing terms are 81 and 243.

Exercise

1 Write down the next three terms in each sequence:

a) 10, 11, 12, …
b) 80, 72, 64, …
c) 1, 2, 4, 8, …
d) 54, 8, 56, 6, 58, 4, …
e) 1, 1, 2, 3, 5, …
f) 1, 0, 0, 1, 0, …

2 Copy these sequences and fill in the missing terms.

a) 3, 10, 17, 24, ☐, ☐
b) 1, 2, 4, 8, 16, ☐, ☐
c) 5, 10, 15, ☐, ☐, 30
d) 42, 38, 34, 30, 26, ☐, ☐
e) 4 096, 1 024, 256, 64, ☐, ☐
f) 3, 5, 8, 10, 13, ☐, 18, 20, ☐
g) 1, 3, 6, 18, 36, ☐, ☐
h) 2, 7, 5, 10, 8, 13, ☐, 16, 14, ☐
i) 0, 25, 50, 75, ☐, ☐, ☐
j) 1, 1, 2, 3, 5, 8, ☐, ☐, ☐

3 Sarah has just opened a sandwich shop. On the first day she has no customers and she sells no sandwiches. On the second day she sells three sandwiches and on the third day she sells 6 sandwiches.

If this pattern continues, how many sandwiches will Sarah sell on:

a) the fourth day **b)** the fifth day **c)** the tenth day?

4 To make one beach-hut out of matchsticks we used six matchsticks. To make two beach-huts we used eleven matchsticks.

a) Draw a diagram showing how you would make three beach-huts.

b) How many matches would you need to make three beach-huts?

c) Without drawing, predict how many matchsticks you would need for four beach-huts.

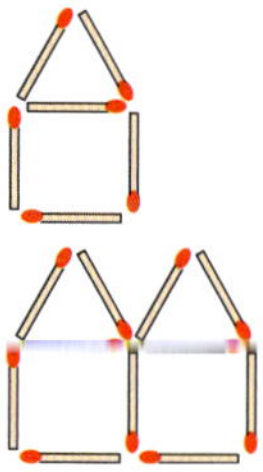

d) Copy and complete this table:

Number of beach-huts	1	2	3	4	5	6	7
Number of matchsticks	6	11					

5 The following number sequence is called Pascal's triangle. Discuss with a partner how it works and then complete the fifth and sixth rows.

1
1 2 1
1 3 3 1
1 4 6 4 1
1   1

6 In the UK, on average days in August have 15 hours of daylight. Days in September have 13 hours of daylight and in October they have 11.

a) How many hours of daylight per day do you think there are on average in November?

b) Why can't this sequence continue using a set pattern?

Investigation

1 The term-to-term rule for a sequence is 'add three'. Using this rule:

a) Write down two possible sequences that follow this rule.

b) Generate a sequence where all the numbers are multiples of 3.

c) Generate a sequence where all the numbers are greater than 100.

d) Generate a sequence where none of the numbers are positive integers.

e) Repeat these activities adding a number other than three. Write down any patterns that you notice.

2 Mark finds the following code written in a maths book. Can you work out what the next term will be?

O, T, T, F, F, S, S, E, N, …

Key words
Factor
Prime factor
Highest common factor

Unit 2.5 Factors

In this unit you will:

- find the factors of a given number
- express a given number as a product of its prime factors
- find the highest common factor (HCF) of a set of numbers
- use HCF to solve problems.

Factors of a number are the numbers that will divide exactly into that number. Factors of 20 are 1, 2, 4, 5, 10 and 20.

Prime factors are those factors which are also prime numbers. Prime factors of 20 are 2 and 5.

Every composite number can be expressed as a product of its prime factors.
$20 = 2 \times 2 \times 5$

To find these prime factors we use division.

- Divide the number by its smallest prime factor. $20 \div 2 = 10$
- Divide the result by its smallest prime factor. $10 \div 2 = 5$
- Continue until the result itself is a prime number. In this example, 5 is a prime factor. $5 \div 5 = 1$.
- Write the product of primes in ascending order. $2 \times 2 \times 5$.

Worked examples

1. List the factors of 24.
2. Express 24 as a product of its prime factors.

1. The factors of 24 are: 1, 2, 3, 4, 6, 8, 12 and 24.
2. $2\overline{)24}$
   $2\overline{)12}$
   $2\overline{)\ 6}$
   $3\overline{)\ 3}$
   $\quad 1$
   $24 = 2 \times 2 \times 2 \times 3$

Exercise

1 Find the factors of each of the following numbers and then express each one as a product of its prime factors.

a) 12	**b)** 18	**c)** 27
d) 32	**e)** 45	**f)** 48
g) 65	**h)** 55	**i)** 72

The highest common factor (HCF)

The **highest common factor** of a set of numbers is the largest factor which is a factor of each number. You can find the HCF of a set of numbers by:

- listing all the factors of each number and identifying the highest number which is common to all of them, or
- writing the numbers as products of their prime factors. The HCF is the product of the prime factors shared by each number.

Worked example

Find the HCF of:

a. 24 and 32 **b.** 18, 36 and 42

Some texts use GCF, or Greatest Common Factor, rather than HCF.

Method 1

a. Factors of 24: 1, 2, 3, 4, 6, **8**, 12, 24 Factors of 32: 1, 2, 4, **8**, 16, 32 The HCF of 24 and 32 is 8.	**b.** Factors of 18: 1, 2, 3, **6**, 9, 18 Factors of 36: 1, 2, 3, 4, **6**, 9, 12, 18, 36 Factors of 42: 1, 2, 3, **6**, 7, 14, 21, 42 The HCF of 18, 36 and 42 is 6.

Method 2

a. $24 = \mathbf{2 \times 2 \times 2} \times 3$ $32 = \mathbf{2 \times 2 \times 2} \times 2 \times 2$ $\text{HCF} = 2 \times 2 \times 2 = 8$	**b.** $18 = \mathbf{2 \times 3} \times 3$ $36 = 2 \times \mathbf{2 \times 3} \times 3$ $42 = \mathbf{2 \times 3} \times 7$ $\text{HCF} = 2 \times 3 = 6$

Exercise

1. Find the HCF of the following sets of numbers.

   a) 14 and 21 **b)** 16 and 18 **c)** 42 and 54

   d) 12, 27 and 30 **e)** 48, 64 and 96 **f)** 13 and 17

2. Ms Clara has three pieces of lace with lengths 48 cm, 45 cm and 54 cm respectively. Ms Clara wants to cut each length into smaller pieces of the same length so that no lace is left over. What is the greatest length that each of the smaller pieces can be?

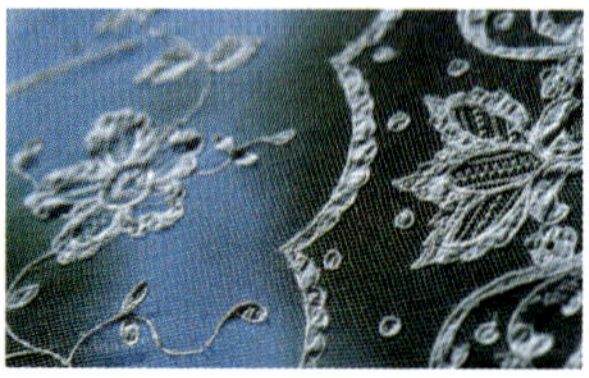

3. Find the largest number which, when divided into 104 and 127, will leave a remainder of 12 in each case.

Investigation

List all the numbers below 50 that have an odd number of factors.
Can you spot a pattern?

What do we call these numbers?

Key words
Multiple
Infinite
Lowest common multiple

Unit 2.6 Multiples

In this unit you will:

- list multiples of a given number
- determine the lowest common multiple (LCM) of a set of numbers
- use LCM to solve problems.

Multiples of a number are that number multiplied by any other whole number. Multiples of 3 are 3, 6, 9, 12, 15, 18 …

Multiples of a number are **infinite**.

We can also say that a multiple of any number n is any number which is exactly divisible by n.

The **lowest common multiple** (LCM) of a set of numbers is the smallest multiple which is common to all sets.

To find the LCM of a set of numbers:

- list the multiples of each number in order and identify the smallest number common to all sets, or
- write each number as a product of prime factors. For each prime factor, choose its greatest product from one of the numbers. The LCM is the product of these prime factors.

Worked example

Find the LCM of 16, 18 and 24.

Method 1

Multiples of 16: 16, 32, 48, 64, 80, 96, 112, 128, **144**, 160, …

Multiples of 18: 18, 36, 54, 72, 90, 108, 126, **144**, 162, …

Multiples of 24: 24, 48, 72, 96, 120, **144**, 168, …

The LCM of 16, 18 and 24 is 144.

Method 2

$16 = \mathbf{2 \times 2 \times 2 \times 2}$ (the largest group of 2s)

$18 = 2 \times \mathbf{3 \times 3}$ (the largest group of 3s)

$24 = 2 \times 2 \times 2 \times 3$

$2 \times 2 \times 2 \times 2 \times 3 \times 3 = 144$

The LCM of 16, 18 and 24 is 144.

Exercise

1. Find the LCM of the following sets of numbers.
   a) 18 and 27 **b)** 6, 8 and 10 **c)** 12 and 22
   d) 14, 21 and 49 **e)** 25, 30 and 36 **f)** 10 and 100

2. Martha goes to church every other day. Junior goes every third day and Jason goes every fourth day. If they are all together at church on Sunday this week, when will they next be at church together?

3. Find the least number of marbles which can be put into bags of 9 or 12 or 15 marbles with none left over.

Unit 2.7 Ratio and proportion

Key words
Ratio
Proportion
Simplify

In this unit you will:

- recognise ratios and write them correctly
- simplify ratios
- solve problems involving ratio and proportion.

Ratio and **proportion** are used to compare quantities.

A proportion compares one part with the whole. For example, three of these four jars are filled with jam.

A ratio is a way of comparing one part with another part. For example, there are 3 jars of jam to every one jar of marmalade.

We can write the ratio of jam to marmalade using the symbols 3 : 1.

We can write the ratio of marmalade to jam using the symbols 1 : 3.

We read '3 : 1' as 'the ratio of three parts to one part'.

We can **simplify** ratios by dividing by a common factor. A ratio of 4 : 8, for example, can be simplified to 1 : 2.

Worked examples

For this grid, write:

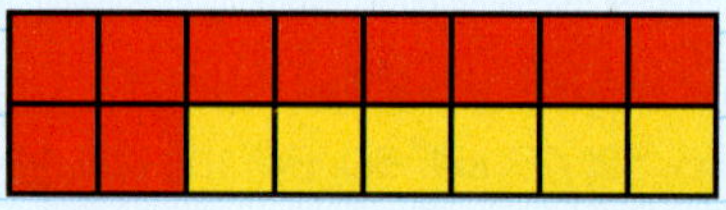

a. the proportion of red squares
b. the proportion of yellow squares
c. the ratio of red to yellow squares
d. the ratio of yellow to red squares

a. 10 : 16 = 5 : 8 **b.** 6 : 16 = 3 : 8 **c.** 10 : 6 = 5 : 3 **d.** 6 : 10 = 3 : 5

Exercise

1 For each grid, write:

a) the proportion of red squares
b) the proportion of yellow squares
c) the ratio of red to yellow squares
d) the ratio of yellow to red squares

i)

ii)

iii)

iv)

2 Copy each grid. Colour the blocks red and yellow to match the given ratios of red to yellow blocks.

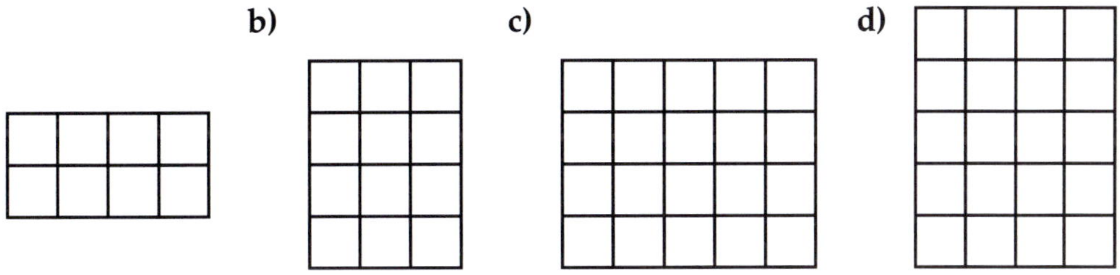

a) 1:3 b) 1:2 c) 4:1 d) 7:3

3 Simplify the following ratios:

a) 6 : 2 **b)** 4 : 20 **c)** 25 : 10 **d)** 12 : 8 **e)** 140 : 210

f) 12 : 120 **g)** 2 : 18 **h)** 16 : 16 **i)** 75 : 35 **j)** 125 : 225

4 Copy this chart, which shows the hours in a day from midnight through to midnight.

	12	1	2	3	4	5	6	7	8	9	10	11	12
a.m.													
p.m.													

Write 'S' for sleeping in each box to indicate your sleeping hours for an ordinary weekday.

Write 'Sc' for hours you normally spend in school.

Write abbreviations for the other hours, for example 'H' for homework, 'O' for out, 'TV' for watching TV, and so on.

a) Write the ratio of hours spent sleeping to hours spent awake.

b) Choose five other ratios to describe your day, for example the ratio of homework to time spent in school.

5 In a café, the proportion of female customers is 3 out of 7. Write the ratio of female to male customers.

6 In a class, 6 out of 30 pupils travel to school by car. Write the ratio in simplest form of those who come by car to those who don't.

7 A school cricket team won 15 and lost 5 of their home matches, and won 7 and lost 13 of their away matches. Write the ratio of:

a) home matches lost to home matches won

b) away matches won to away matches lost

c) total matches won to total matches lost.

Investigation

Investigate the ratio of the number of prime numbers to the number of non-prime numbers in:

a) the numbers from 1 to 50 inclusive **b)** the numbers from 1 to 100 inclusive.

Unit 2.8 Problems involving ratio

Key words
Ratio

In this unit you will:

- divide an amount into parts using a given ratio
- solve problems involving ratio.

To divide an amount in a given **ratio**, there are four stages:

For example, to divide \$15 in the ratio 2 : 3.

- Find the total number of small parts. Two small parts and three small parts is a total of five small parts.
- Work out how much each small part is worth. \$15 divided by 5 is \$3.
- Multiply each part of the ratio by this amount. 2 × \$3 = \$6 and 3 × \$3 = 9
- Check the two parts are in the given ratio and that the total is correct: (\$6 : \$9 = 2 : 3 and \$6 + \$9 = \$15).

Worked example

Share 60 plums among Ken, Bernie and Daria in the ratio 2 : 3 : 5 respectively.

Total number of small parts = 2 + 3 + 5 = 10

One part = 60 ÷ 10 plums = 6 plums

Ken: 6 plums x 2 = 12 plums

Bernie: 6 plums x 3 = 18 plums

Daria: 6 plums x 5 = 30 plums

Exercise

1. Divide \$24 in the following ratios:

   a) 1 : 2 **b)** 3 : 5 **c)** 3 : 1

   d) 7 : 5 **e)** 1 : 7 **f)** 19 : 5

2. Sally and Jonathan have earned \$36 cleaning cars. They divide the money according to how much work they have each done. How much will they each receive if they decide to divide it in the following ratios?

   a) 5 : 1 **b)** 4 : 5 **c)** 1 : 3

   d) 11 : 1 **e)** 5 : 13 **f)** 1 : 1

3. There are 2 boys to every 5 girls at an outing. If there are 35 girls, how many boys are there?

4. Share \$112.00 among Anna, Fred and Matt in the ratio 1 : 2 : 3.

5 A cake recipe uses 8 eggs for every 1 kilogram of butter.

a) How many eggs should be used for:

i) 3 kilograms of butter

ii) 4 kilograms of butter

iii) 2 kilograms of butter

iv) 5 kilograms of butter?

b) How many kilograms of butter should a caterer use with:

i) 64 eggs **ii)** 96 eggs

iii) 48 eggs **iv)** 104 eggs?

6 A seamstress makes 7 dresses in 3 days. If she works at the same rate, how many of these dresses will she make in:

a) 6 days **b)** 27 days **c)** 3 weeks?

7 36 baseball cards were shared between Harry and Tom. Harry received twice as many as Tom. How many baseball cards did each boy receive?

8 Share \$80.00 among Bruce, Eric and Babsie in the ratio 4 : 5 : 7 respectively.

Case study - Oliver Joseph

Mr Joseph has been transporting students to secondary schools for the past 10 years. Last year he had two 20-seater buses and each bus carried 18 students on each trip to St James Secondary School. Each bus made one trip on a school day.

The Ministry of Education recently announced that there would be more places available in secondary schools next year and so more students will be able to attend. As a result, it was discovered that there would be 300 students from Mr Joseph's area attending secondary schools next year.

The news made Mr Joseph excited, and he is contemplating selling his two buses and buying a bigger one which would be able to transport 50 students. He is also aware that there are other bus drivers in the area who will be transporting students to secondary school. The following table shows these bus drivers, the number of buses owned and the number of students that the drivers have been contracted to transport.

Bus driver	Number of buses	Number of students per bus
Mr Tom	1	20
Mr Casey	2	38
Mr George	4	17
Mr Stan	3	18
Mr Haynes	3	19

Mr Joseph is not very good with numbers. He needs to decide whether to buy the bigger bus and get rid of the smaller ones. Advise Mr Joseph, giving reasons for your decision.

Unit 2.9 Solving problems

In this unit you will:

- develop different strategies for solving problems
- apply your skills to solve a range of word and picture problems.

Key words
Solution
Trial and error
Patterns
Rules

There are many different types of problems in mathematics and there are many different techniques that you can use to solve them. Two different approaches to problem solving are shown below:

Stage 1	Stage 2	Stage 3
Reading the problem • Do I understand what it means? • Am I sure that the information is clear?	Information that is given • What do we know for sure? • Is there anything else we can work out?	Thinking about a **solution** • What are we trying to find? • What will help us? • What is the best method to find a solution?

Make a plan.
Choose a technique and organise the information.

↓

Try out your plan.
Use your technique to get an answer.

↓

Check your solution.
Is your answer realistic and logical? Compare with others.

Discuss with a partner how these two plans can help you to solve problems.

Logical thinking, and trial and error

How far can a parrot fly into a forest?

Zara and Ishmael try to answer this question. Ishmael says it cannot be solved because they don't have enough information. He says they need the size of the forest. Do you agree?

Zara thinks for a while and then she laughs and says: 'It can be answered. The parrot can only fly halfway into the forest. After that, it is flying out of the forest!'

Zara used her thinking skills to solve this trick problem.

Some problems can be solved by trying out solutions till the right answer is reached. This is called **trial and error**.

Worked example

A farmer is looking at his animals. There are goats and chickens in the yard. Altogether there are 18 heads and 56 legs. How many goats are there?

$G + C = 18$ (because there are 18 heads there must be 18 animals altogether)

9 goats = 36 legs; 9 chickens = 18 legs 36 + 18 = 54 legs ✗

10 goats = 40 legs; 8 chickens = 16 legs 40 + 16 = 56 legs ✓

So there must be 10 goats.

Exercise

Use thinking skills and trial and error techniques to find the solutions to these problems. In each one, write down what technique you used.

1 Copy the grid into your book. Use each number from the box once only to fill the missing spaces so that the numbers add up to 100.

3 6 2 5 7 1 4

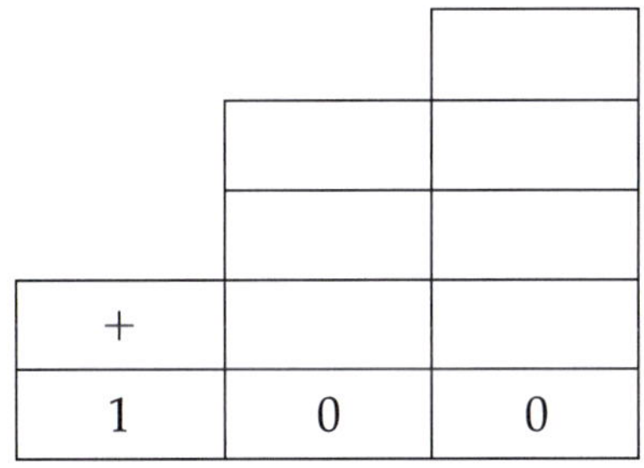

2 The calculations below are alphanumeric. Each letter represents a specific digit. Try to replace the letters with digits to find solutions that work. Use the clues to help you.

a)

```
  N I N E
– F O U R
  F I V E
```

Clues: Try R = 0, N = 5
There are more than 60 solutions!
Try to find five.

b)

```
  T H R E E
+   F O U R
  S E V E N
```

Clues: Try E = 6, V = 0 or E = 5, V = 1
There are more than 30 solutions.
How many can your class find?

3 How many triangles are there in each of these shapes?

a)

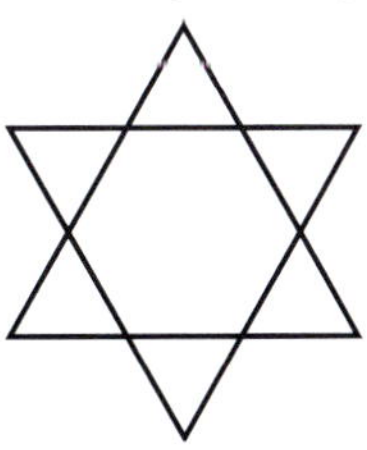

b)

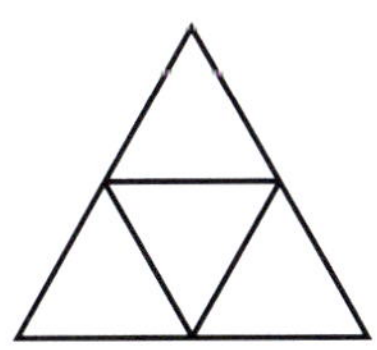

c)

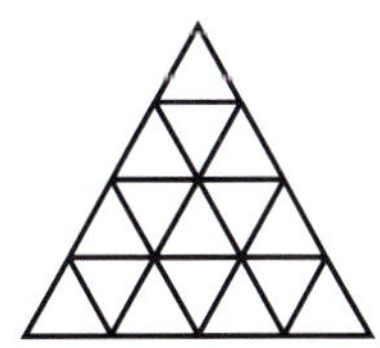

Using patterns and making rules

Some problems can be solved by looking for **patterns**. Once you have found a pattern, you can use it to make a general **rule**. For example:

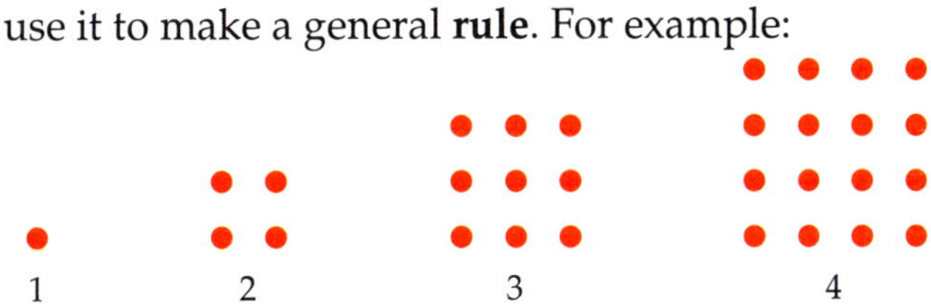

The sides of the squares contain the same number of dots as the position. The first square has one dot per side, the second has two and so on.

Worked example

How many dots will you need to make the 10th square in the pattern above?

Square 1:	1 x 1	= 1
Square 2:	2 x 2	= 4
Square 3:	3 x 3	= 9
Square 4:	4 x 4	= 16
So square 10:	10 x 10	= 100 dots

Exercise

Use patterns and make rules to find the solutions to these problems.

1 Look at the matches in the patterns below.

a) What is the perimeter (in match lengths) with 20 triangles?

b) What is the perimeter (in match lengths) with 100 triangles?

c) Write a rule for the pattern.

2 What are the next two numbers in these patterns? Write down the rule that you used in each case.

a) 2, 1, 4, 3, 6, 5

b) 1, 1, 2, 3, 5, 8

c) 1, 8, 27, 64

d) 2, 5, 8, 11

3 John builds these patterns with matches:

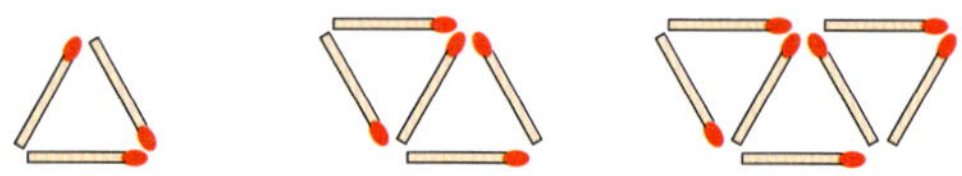

a) How many matches will he need to build 7 triangles?

b) How many matches will he need to build 100 triangles?

c) Write a rule that tells how many matches you will need to build any number of triangles.

Diagrams and tables

Sometimes a diagram can be used to show a problem more clearly. Think back to the goats and chickens problem on page 28. We can draw it like this:

18 heads with 2 legs each = 36 legs
20 legs left so give 2 to each goat

So 10 goats

A table can also help you to see patterns and rules more clearly. Think back to the perimeter of triangles problem in question 1 above.

No. of triangles	1	2	3	4	?
Perimeter in match lengths					

Looking back

1. The table shows the total number of secondary school students in five districts.

District	Number of students
Caroni	3 912
South Eastern	4 875
Tobago	3 002
St Patrick	6 135
North Eastern	2 879

   a) What is the difference between the numbers of secondary school students in Tobago and North Eastern?

   b) What is the total number of secondary school students in all five districts?

   c) Which district has about twice the number of secondary school students as Tobago?

2. Do the following calculations:

   a) $31 + (21 - 15) \times 14$ b) $11 + 9 \times 4 \div 6 - 5$ c) $(3 \times 5 + 8) - (45 \div 5) \times 2$

3. Are subtraction and division commutative operations? Give an example to support your answer.

4. Complete the following sequences:

   a) 1, 10, 19, 28, □, □ b) 1, 2, 4, 7, 11, □, □ c) 1, 5, 25, □, □

5. Fill in the blanks:

   a) 1 : 2 = 15 : □ b) 6 : 9 = □ : 3 c) 5 : 16 = □ : 80

   d) 11 : 4 = □ : 48 e) 7 : 3 = 42 : □

6. Simplify the following ratios:

   a) 18 : 21 b) 35 : 49 c) 24 : 108

   d) 90 : 36 e) 22 : 55 f) 26 : 65

7. A carpenter makes 5 tables in 3 days. If he works at the same rate,

   a) how many days will he take to make:

      i) 10 tables ii) 45 tables?

   b) how many tables can he make in:

      i) 18 days ii) 42 days?

8. The ratio of children to adults at a fair is 9 : 4. There are 207 children at the fair. How many adults are there?

9. The ratio of males to females at a jazz concert is 7 : 8. There are 1 200 females at the concert. How many people are there at the concert altogether?

10. Draw a diagram to help you solve this problem.
    There are five people in a queue at the post office. Which are men and which are women?

    - The first person is not Mrs Smith.
    - The last person is a man.
    - There are two men in the queue.
    - The third and fourth people are the same gender.
    - The second person is the brother of the first.

Chapter 3

Sets

In this chapter, you will learn the concept of a set and the terms and symbols associated with sets. You will also learn about Venn diagrams and use them to solve problems.

Key words
Set
Members
Elements
Equal
Equivalent
Finite
Infinite
Empty set

Unit 3.1 The language of sets

In this unit you will:

- define sets and list the elements of sets
- differentiate between equal and equivalent sets and finite and infinite sets
- define empty sets
- use symbols related to sets.

A **set** is a well-defined collection of items. For example, {1, 2, 3} or the set of vowels {a, e, i, o, u}. Curly brackets { } are used to define or contain sets.

The items belonging to a set are called the **elements** of the set. The symbol $\in$ means 'is an element of'. For example, a $\in$ {a, e, i, o, u}.

The symbol $\notin$ means 'is not an element of'. For example, n $\notin$ {a, e, i, o, u}.

Equal sets have exactly the same elements. For example, {1, 2, 3} = {2, 3, 1}.

Equivalent sets have the same number of elements, but the elements themselves do not have to be the same. We use the symbol $\simeq$ to show that sets are equivalent. For example, {1, 2, 3, 4} $\simeq$ {a, b, c, d}.

Finite sets have a definite number of **members** and a first and last element. For example, {1, 2, 3, 4, 5} is a finite set with five elements.

A set is **infinite** if we cannot write down all the members. For example, the set of counting numbers above 10 is infinite. We use … to show that elements of a set continue to infinity. For example, {1, 2, 3, 4, …}.

The **empty set**, or null set, has no elements. For example, {cows with wings}. The empty set is denoted by { } or $\varnothing$.

Exercise

1. Give two examples of:
   a) a well-defined set
   b) a collection of arbitrary objects which is not well-defined.
   c) Is the collection in part **b)** a set? Explain why or why not.
2. Describe the following sets in words:
   a) {a, b, c, d, e}
   b) {cats, dogs, birds, fish}
   c) {Barbados, St Lucia, St Vincent, Jamaica, Cuba}
   d) {January, February, March, April}
   e) {12, 14, 16, 18, 20 …}

3 Match each set in Column A to its equivalent set in Column B. Which of the sets are equal?

Column A	Column B
{5, 6, 7}	{3, 2}
{2, 3}	{14, 12, 11, 9, 10}
{cat, dog, pig, bird}	{6}
{10, 11, 12, 13, 14}	{0, 1, 2, 3, 4, 5, 6}
{whole numbers from 1 to 6}	
{a}	{c, a, t}
{ }	{2, 4, 6, 8}
{3, 5, 7, 9, 11, 13, 15}	{whole numbers from 20 to 25}

4 List two equivalent sets with four members.

5 List the members of the following sets:

a) prime numbers between 1 and 10

b) days of the week

c) the Windward Islands

d) even numbers between 5 and 15

e) multiples of 3 between 10 and 25

6 Fill in ∈ or ∉ in the following sentences.

a) blue ☐ {yellow, red, black, green}

b) e ☐ { vowels}

c) Barbados ☐ {St Lucia, Guadeloupe, Barbados, Jamaica}

d) 2 ☐ {whole numbers divisible by 4}

e) chair ☐ {furniture}

7 State whether each set is finite or infinite.

a) the set of whole numbers

b) the set of letters in your name

c) the set of natural numbers

d) the set of even numbers

e) the set of consonants

f) the set of odd numbers

g) the set of letters in the word MATHEMATICS

h) the set of members of your family

i) the set of subjects you study at school

j) the set of students at your school

8 Which of the following are empty sets?

a) odd numbers divisible by 2

b) factors of 24 between 20 and 30

c) months of the year beginning with W

d) whole numbers between 5 and 6

e) prime numbers which are even

Unit 3.2 Subsets

In this unit you will:

- list the subsets of a set
- use the symbols ⊂ and ⊄ correctly.

Key words
Set
Subset
Proper subset
Improper subset
Universal set

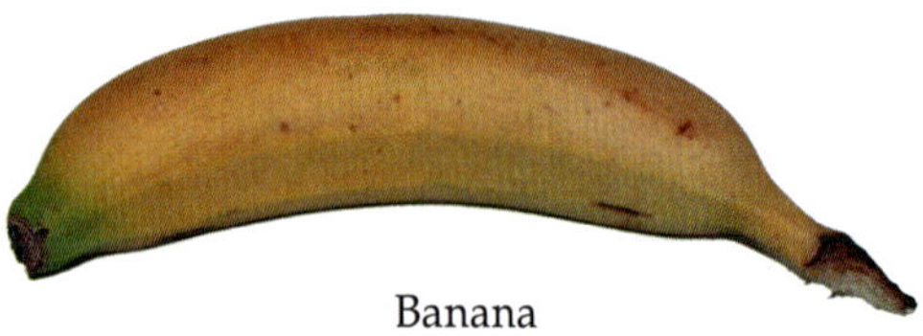

Banana

Orange

Mango

This is a **set** of fruit.

A **subset** is a set within a set. If every element of set A is also an element of set B, then A is a subset of B. This is written as: $A \subset B$.

$\subset$ means 'is a subset of'.
$\not\subset$ means 'is not a subset of'.

A **proper subset** contains at least one element less than the set of which it is a subset. The proper subsets of the set of fruit above are:

{mango} {banana} {orange}

{mango, banana} {mango, orange} {banana, orange}

The empty set, { }, is a subset of every set.
Each set is also a subset of itself. We call it an **improper subset**.

The set of elements from which we select elements to form subsets is called the **universal set** and is normally shown by the letter U.

Exercise

1. When is set A a proper subset of set M?

2. **a)** Write down all the subsets of P = {a, b, c}.
   b) Which of these are proper subsets?

3. Fill in $\subset$ or $\not\subset$:
   a) {yellow, red} ☐ {colours of the rainbow}
   b) {Jupiter, Pluto, Mars, Moon} ☐ {planets of the Solar System}
   c) {whale, dog, cat} ☐ {four-legged mammals}
   d) {chair, table, desk} ☐ {furniture}
   e) { } ☐ {birds}

4. List all the subsets of
   a) { }
   b) {table}
   c) {pencil, pen, eraser}
   d) {2, 3, 4, 5}
   e) {penny, dollar}

5. a) List the set of members of your immediate family.
   b) Write down four subsets of that set.

6. a) List five members of the set of your favourite foods.
   b) Write eight subsets of that set.

7. If X = {1, 3, 5, 7, 9, 11}, list the following subsets.
   a) a subset of X which contains only prime numbers
   b) a subset of X which contains all the multiples of 3
   c) a subset of X which contains only multiples of 4

8. Give a universal set for each of the following proper subsets:
   a) {2, 4, 6, 8}
   b) {33, 44, 55 …}
   c) {Nassau, Castries, Bridgetown}
   d) {Africa, Asia, Australia}

Investigation

1. What is the difference between $A \subset B$ and $B \subset B$?

2. a) How many subsets are in:
      - the empty set
      - a set with one element
      - a set with two elements
      - a set with three elements?
   b) Find the number of subsets in a set with five elements without writing all the subsets.
   c) How many subsets would a set with n elements have?

Unit 3.3 Venn diagrams

In this unit you will:

- draw and interpret simple Venn diagrams.

Key words
Universal set
Venn diagram

We can represent the **universal set** and its subsets using **Venn diagrams**. In the Venn diagram below, U = {whole numbers from 1 to 10}.

P is a subset of U. We can say U contains P. For this reason, we show P inside U but we do not repeat the elements.

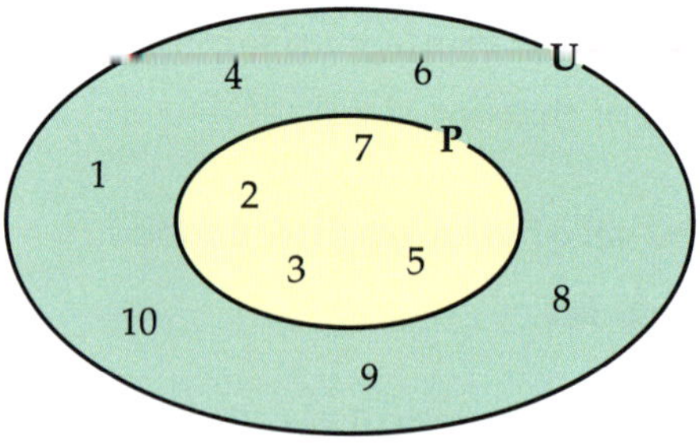

Within this example, P is the set of prime numbers. Because $P \subset U$, P can also be described as the set of prime numbers from 1 to 10.

Exercise

1. Describe U, A and B in words for the diagrams below.

a)
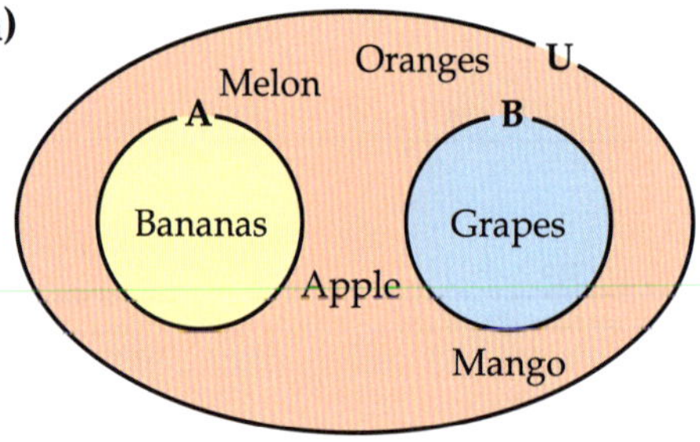

b)
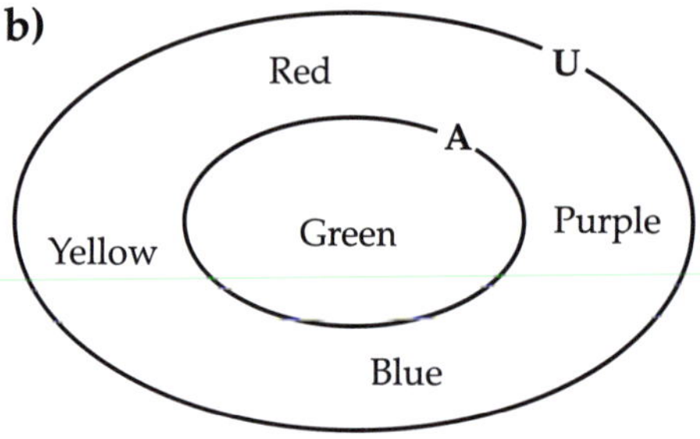

c)
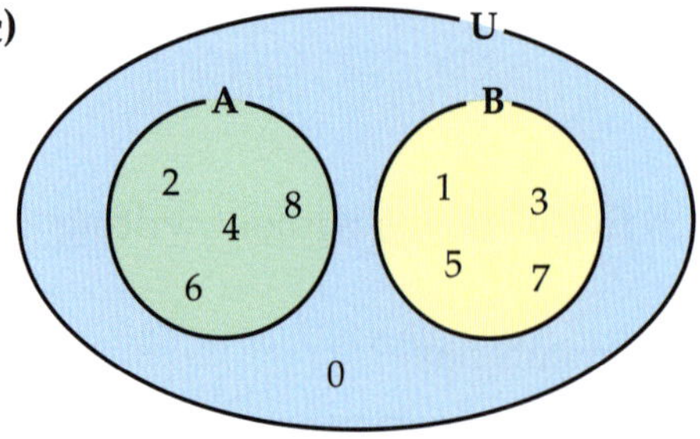

d)
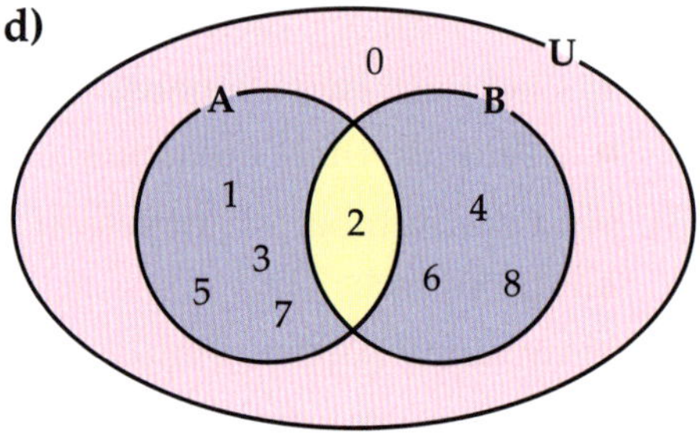

2 Draw Venn diagrams to illustrate the following:

	Universal set	Subset
a	Whole numbers from 1 to 10	Odd numbers from 1 to 10
b	Whole numbers from 6 to 20	Even numbers from 6 to 20
c	Whole numbers from 30 to 50	Composite numbers from 30 to 50
d	Alphabet	Vowels
e	Months of the year	Months with 30 days
f	Days of the week	Days beginning with T
g	Even numbers from 2 to 20	Factors of 16 from 2 to 20
h	Whole numbers from 0 to 20	Prime numbers from 0 to 20
i	Alphabet	Letters in your name
j	Members of your family	Children in your family

3 Represent the following as Venn diagrams:

a) U = {odd numbers from 1 to 20}; T = {multiples of 3 below 20}.

b) U = {whole numbers from 10 to 30}; H = {factors of 100 between 10 and 30}.

4 Use the Venn diagram to find the answers to the following problems:

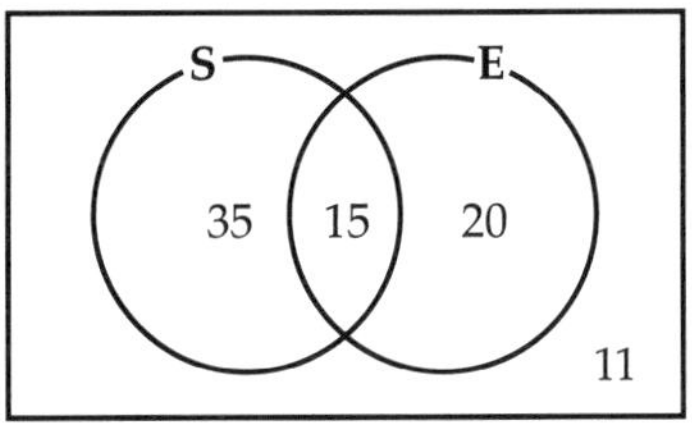

In a group of people there are Spanish-speaking and English-speaking people. There are some people who speak both Spanish and English, and some who cannot speak either language.

a) How many people can speak Spanish?

b) How many can speak English?

c) How many people speak both Spanish and English?

d) How many people speak neither Spanish or English?

5 Represent this information on a Venn diagram:

a) In a class of 36 pupils, 5 take physics, 9 take chemistry and 12 take both. The rest of the pupils take neither of these subjects.

b) In a group of 60 students, 40 play basketball, 30 play cricket and 8 play neither game.

Looking back

1. List the members of the following sets.
   a) letters in the word CARIBBEAN
   b) factors of 36 between 0 and 10
   c) five students in your class
   d) days of the week beginning with J
   e) composite numbers between 19 and 29

2. Give two examples of a finite set and two examples of an infinite set.

3. Which of the following are empty sets?
   a) multiples of 6 which are not multiples of 3
   b) whole numbers between 1 and 2
   c) prime numbers divisible by 5

4. a) List all the subsets of {1, 2, 3, 4}.
   b) How many are there?

5. Show this as a Venn diagram:
   U = {months of the year}; M = {months with 31 days}

6. Draw Venn diagrams to illustrate the following.

	Universal set	**Subset**
a	Even numbers from 1 to 30	Numbers less than 30 divisible by 4
b	Alphabet	Letters in the word EXTRAORDINARILY

Revise and consolidate

1. Write these numbers in words:
   - **a)** 4 732
   - **b)** 800 073
   - **c)** 3 009
   - **d)** 1 301 052

2. What is the value of the 3 in each of the numbers in question 1?

3. Complete these calculations.
   - **a)** $5 \times 4 + 2 \times 3$
   - **b)** $9 \times 2 + 3 \times 2$
   - **c)** $6 \times 2 + 5 \times 3$
   - **d)** $8 \times 4 + 15 \div 3$
   - **e)** $12 \div 6 - 10 \div 5$
   - **f)** $7 \times 2 - 18 \div 3$
   - **g)** $24 \div 2 + 4 \times 4$
   - **h)** $18 \div 1 - 3 \times 4$
   - **i)** $7 \times 5 - 4 \times 6$
   - **j)** $3 \times 8 + 40 \div 5$
   - **k)** $60 \div 10 - 40 \div 5$
   - **l)** $9 \div 9 + 4 \times 7$

4. Fill in the missing operation signs to make the following correct.
   - **a)** $5 - 4 + 1 = 5 \square 1 \square 4$
   - **b)** $8 - 6 + 4 - 2 = 8 \square 2 \square 4 \square 6$
   - **c)** $19 + 7 - 12 + 5 = 19 \square 12 \square 5 \square 7$
   - **d)** $12 \times 3 \div 2 = 12 \square 2 \square 3$
   - **e)** $18 \div 3 \times 6 = 18 \square 6 \square 3$
   - **f)** $120 \div 6 \times 2 \div 10 = 120 \square 10 \square 6 \square 2$

5. Find numbers which satisfy the following descriptions.
   - **a)** a prime number which is a factor of 17
   - **b)** a number whose factors include 3 and 9
   - **c)** a number which has only two factors
   - **d)** a prime number which has an even factor
   - **e)** a number with only one factor
   - **f)** a number which has three factors only

6. Find the HCF of the following sets of numbers:
   - **a)** 32 and 60
   - **b)** 48 and 90
   - **c)** 21, 28 and 42
   - **d)** 24, 36 and 64

7. Find the LCM of the following sets of numbers:
   - **a)** 36 and 48
   - **b)** 18 and 52
   - **c)** 6, 10 and 16
   - **d)** 26, 39 and 52

8. Find the smallest number of sweets that can be put into bags each containing 26 or 39 or 65 sweets with none left over.

9. Find the largest number which, when divided into 163 and 237, will leave a remainder of 15 in each case.

10 Write in index form:

a) $5 \times 5 \times 5 \times 5 \times 5$
b) $3 \times 3 \times 3$
c) $12 \times 12 \times 12$
d) $1 \times 1 \times 1 \times 1 \times 1 \times 1 \times 1$
e) 7×7
f) $3 \times 3 \times 3 \times 3 \times 3 \times 3$
g) $25 \times 25 \times 25 \times 25 \times 25$
h) $10 \times 10 \times 10 \times 10 \times 10 \times 10$
i) 1000×1000
j) $2 \times 2 \times 2 \times 2$

11 Write in expanded form and then calculate:

a) 2^3 **b)** 4^2 **c)** 5^3 **d)** 10^0
e) 8^6 **f)** 10^5 **g)** 9^6 **h)** 15^5

12 Use symbols to represent the following:

a) a set of the first five letters of the alphabet
b) a set of all the odd numbers
c) set C is a subset of set B
d) the empty set
e) c is not an element of a set containing a and b
f) set B is equivalent to set X

13 Illustrate the following using Venn diagrams:

a) The set of all scholars on your island; the set of scholars in your town; the set of scholars in your school.
b) The set of whole numbers from 1 to 9; the set of the first three whole numbers; the set of even numbers from 1 to 9.

14 Describe what is shown by this Venn diagram:

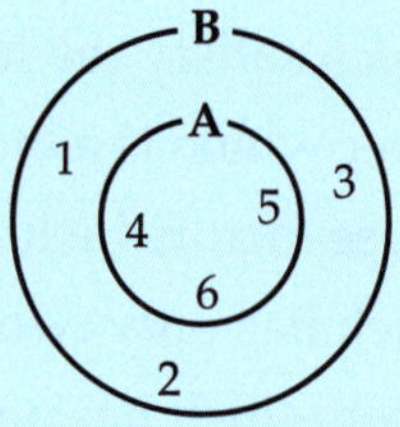

15 If A = {5, 6, 7, 8} and B = {8, 9, 10, 11}:

a) State whether set A is equal to set B and give a reason.
b) State whether set A is equivalent to set B and give a reason.
c) Draw a Venn diagram to illustrate these sets.
d) Write down the elements of set X if X is a subset of A and X is a subset of B.

Key words
Non-standard
Standard

Chapter 4

Measuring systems

People have used different ways of measuring for thousands of years. This chapter looks at the history of measuring systems and tools, and the development of systems which use standard units. You will find out about the imperial and metric systems of measuring and also explore how we measure time and temperature. You will work with units of measure, estimate measurements and express measurements in different forms.

Unit 4.1 Non-standard units

In this unit you will:

- research non-standard units of measure and how they were used long ago
- appreciate the need for standard units of measure.

People have used different systems for measuring over the ages. Some of these systems used parts of the body as their units. The Egyptians used a unit called a cubit – the distance from a man's elbow to the tip of his longest reaching finger. They also used digits (width of a finger), palms (the width of four fingers) and hands (five digits). The Romans introduced the foot and the pace as units of measure.

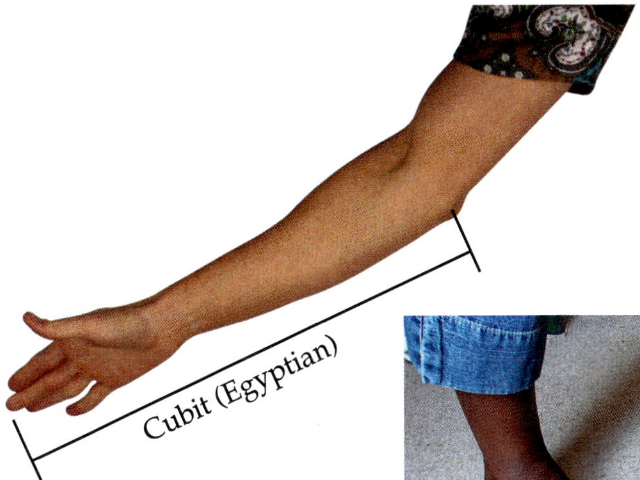

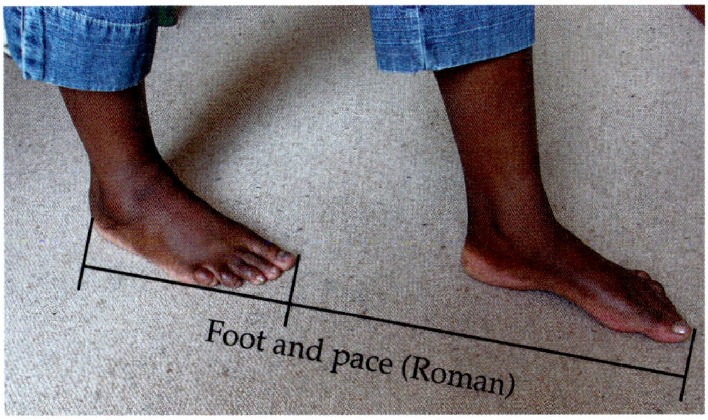

We call these early units **non-standard** units because their value depended on whose body was being used – some people had bigger cubits than others! People eventually agreed that they all needed to use the same units to be fair. In Egypt, the pharaoh kept a standard for the cubit. All other measuring sticks based on the cubit had to be the same length and inspections were carried out to check this. Over time, a number of **standard** units of measure have been developed and the units we use today are the same all over the world.

Exercise

1. Find the meaning of these words and write down what was measured in each unit:

   a) furlong **b)** span **c)** mina **d)** horsepower

2. Work in groups of three. Draw an outline of your foot on a sheet of card. Cut out the foot you have drawn and write your name on it. Choose five objects to measure, and record the measurements on a table like this:

Object to be measured in 'feet'	My measure	Partner 1's measure	Partner 2's measure

3. **a)** What do you notice about the measurements in question 2?

   b) What does this tell you about the need for standard units of length?

   c) Find out the length of a foot in standard measuring units.

4. It's fairly easy to use parts of the body to estimate length. But what about mass? How do you think people in ancient times measured and compared mass?

5. Make a list of five things that are measured in everyday life and the units they are measured in. With a partner, discuss the problems that might arise if we did not have standard units for these measurements.

Investigation

Old textile sellers used to measure cloth lengths by holding the fabric and stretching it along the arm to their nose, like this:

Today we can still use our bodies to estimate lengths in metres.

Try this:

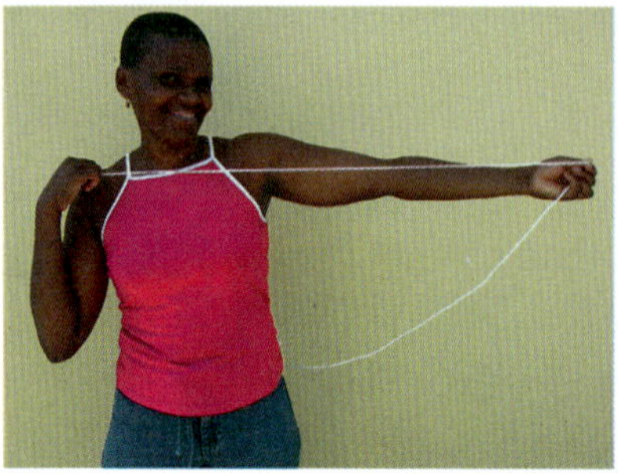

How long is this distance on average?

Carry out your own investigation using string and a tape measure.

Record and compare the results from at least five students.

Who has a perfect metre?

How close are the rest of you?

When could it be useful to be able to estimate a metre like this?

Key words
Imperial
Metric
Length
Mass
Capacity

Unit 4.2 Metric and imperial units

In this unit you will:

- learn the names of imperial and metric units of length and mass
- identify the instruments used to measure different quantities
- examine the relationship between metric and imperial units of measure
- estimate and compare measurements in imperial and metric units
- convert between imperial and metric units.

The British developed a standard system of measurements called the **imperial** system. The imperial system is based on units such as inches, pounds and pints, which are still used in some parts of the world today. The relationship between units in the imperial system makes it clumsy to work with and scientists, engineers and manufacturers in different countries have tended to move towards the **metric** system.

The metric system was developed by the French. The base unit of **length** in this system is the metre. The base unit of **mass** is the kilogram. The base unit of **capacity** is the litre. These are the main units that you will work with this year:

1 kilometre	= 1 000 metres	1 tonne	= 1 000 kilograms
1 metre	= 100 centimetres	1 kilogram	= 1 000 grams
1 centimetre	= 10 millimetres	1 gram	= 1 000 milligrams

Some imperial units and their metric equivalents are shown here:

Length
Imperial: 12 inches is 1 foot;
36 inches or 3 feet is 1 yard.
Metric: 10 millimetres is 1 centimetre;
100 centimetres is 1 metre.

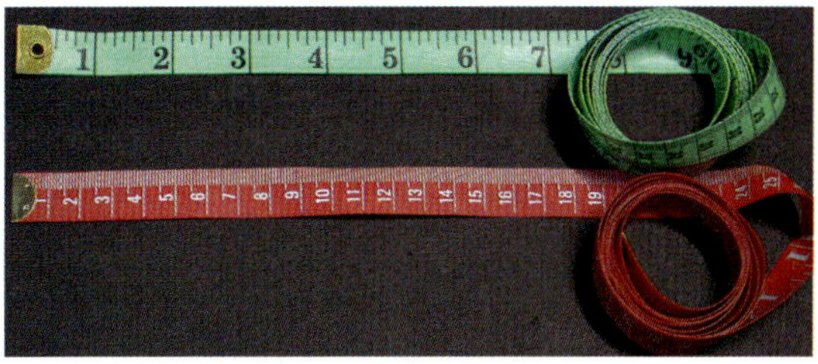

Mass
Imperial: 16 ounces is 1 pound (lb)
Metric: 1 000 grams is 1 kilogram

Capacity
Imperial: 16 fluid ounces is 1 pint.
8 pints is 1 gallon.
Metric: 1 000 millilitres is 1 litre.

Temperature
Imperial: degrees Fahrenheit (°F)
Metric: degrees Celsius (°C)
Water freezes at 32 °F or 0 °C.

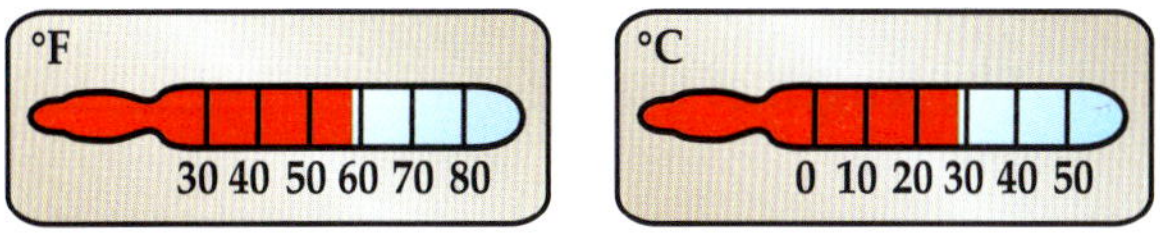

The metric system is widely used in the Caribbean but some islands still use imperial measures, particularly in markets and stores.

Exercise

1. Say which metric unit would be suitable to measure the following:
   a) the length of your foot
   b) the mass of a pencil
   c) the distance from St Lucia to Barbados
   d) your mass
   e) the distance around an athletics track
   f) the thickness of a slice of bread

2. Which measurement is most likely to be accurate in each case?
   a) The distance across a classroom is:
   50 cm   50 m   5 cm   5 m
   b) The length of a lizard's tail is:
   120 mm   120 cm   120 m   120 km
   c) A chocolate has a mass of approximately:
   3 g   1 kg   300 g   3 mg

3. Estimate the following:
   a) The length of this cruise ship.

   b) The mass of the child.
   The adult has a mass of 76 kilograms.

   c) The length of your cubit.
   d) The mass of a teaspoon of sugar.
   e) The length of these lines:

   f) The heights of the tallest and shortest students in your class.

Working with imperial and metric units

This table gives you some quick methods of estimating the metric equivalent of imperial measures.

Imperial	Metric	Approximate	More accurate
Yard (3 ft or 36 in)	Metre	÷1	÷1.094
Foot (12 in)	Metre	÷3.3	÷3.281
Inch	Centimetre	×2.5	×2.54
Ounce (oz)	Gram	×28	×28.349
Pound (lb) (16 oz)	Kilogram	×0.5	×0.454

Worked examples

1. Peter has a pencil that is 3 inches long. Approximately how long is it in centimetres?
2. Jerry is 6 feet tall. Approximately how tall is this in metres?
3. A bag of sand has a mass of 5 kilograms. Approximately how many pounds is this?

1. 3 in x 2.5 = 7.5 cm
   It is approximately 7.5 cm long.
2. 6 ft ÷ 3.3 = 1.82 m
   He is approximately 1.8 m tall.
3. 5 x 2 = 10 pounds.
   It has a mass of approximately 10 pounds.

To convert metric to imperial, use the opposite operation.

Exercise

1. List three things that are about 1 metre long.
2. List three things that are about 1 inch long. About how long are these items in centimetres?
3. Which bottle holds more: a 1 litre bottle or a 1 pint bottle?
4. How tall are you in metres? How tall is this in feet?
5. Trudi is 5 feet 2 inches tall. Her sister is 160 cm tall. Who is taller?
6. A fabric shop sells silk ribbons at $8.00 for 1 metre. Mrs Singh wants to buy 25 yards of ribbon. How much will this cost her?
7. Mr Proudfoot complained that he needed to lose 20 pounds. Approximately how many kilograms does he have to lose?

Unit 4.3 Temperature

In this unit you will:

- convert between Fahrenheit and Celsius temperature measurements
- solve problems involving temperature.

Key words
Temperature
Degrees
Fahrenheit
Celsius

Temperature is a measure of how hot or cold something is. Temperature is measured using a thermometer. The liquid in the thermometer moves up and down the tube and the temperature is read from the scale in **degrees**. This thermometer shows a temperature of 25 °C.

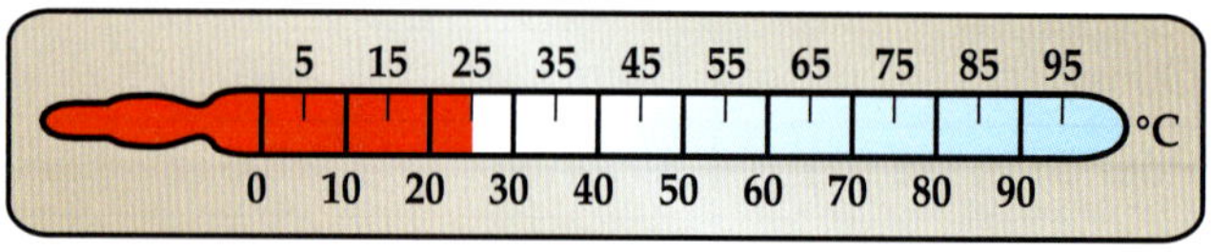

You have already learned about the **Fahrenheit** and **Celsius** scales. When you work with temperature you may need to convert from one to the other. To do this we can use formulae or a graph.

Fahrenheit to Celsius: Subtract 32 and then multiply by $\frac{5}{9}$.
Celsius to Fahrenheit: multiply by $\frac{9}{5}$ and then add 32.

To convert from °C to °F, work in a similar way but using the horizontal scale as a starting point.

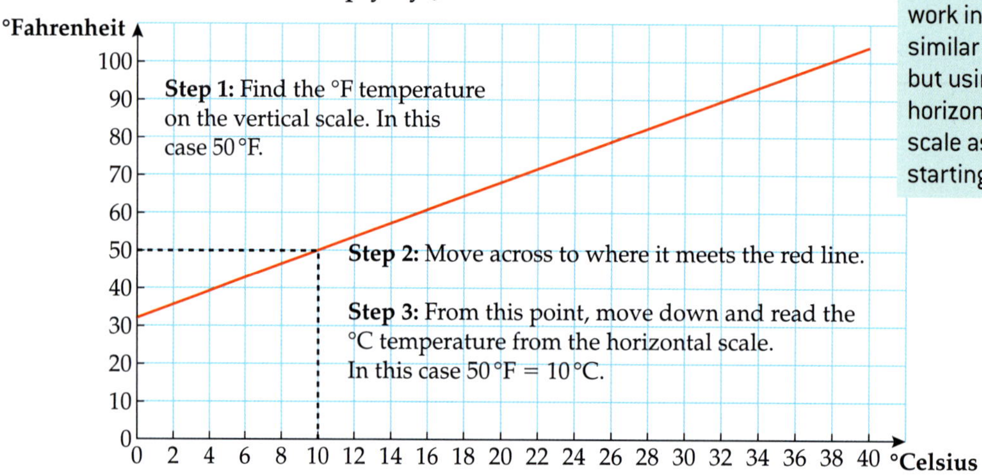

Worked examples

1. Express 104 °F in °C.
2. The temperature in Port-of-Spain is 35 °C. The temperature in New York is 10 degrees cooler.
   a. What is the temperature in New York?
   b. The Americans use the Fahrenheit scale. Use the conversion graph to find the temperature in New York in °F.

1. $104\ °F = (104 - 32) \times \frac{5}{9} = 40\ °C$

2. a. $35\ °C - 10\ °C = 25\ °C$
   b. The temperature in New York is 25 °C. This is equivalent to 77 °F.

Exercise

1 Write down the temperature shown on each thermometer.

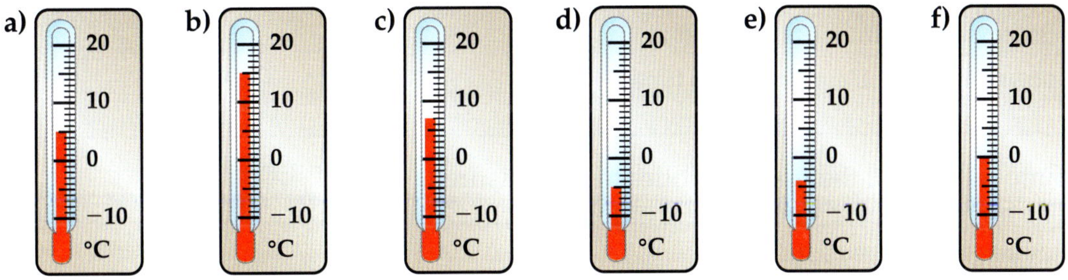

2 Convert these temperatures to degrees Celsius:

a) 50 °F **b)** 71.6 °F **c)** 96.8 °F

3 Convert these temperatures to degrees Fahrenheit:

a) 23 °C **b)** 45 °C **c)** 60 °C

4 Use the graph on page 46 to help you put these temperatures in ascending order:

12 °C	212 °F	90 °C	−12 °C	−50 °F	100 °F	29 °C	50 °F	6 °C

5 The temperature in Bridgetown was 25 °C on Saturday morning. It rose 5 degrees on Sunday.

a) What was the temperature in °C on Sunday?

b) Express this in Fahrenheit.

Case study - Mr Toppin, painter

Mr Toppin works as a painter with a construction firm. His job is to paint the interior and exterior walls, the roof and any other parts of the finished building that may require painting. Mr Toppin enjoys his job, especially creating new and exciting colours and deciding which colours should be used to make the house as beautiful as possible. Sometimes he even paints designs on the walls, like birds, flowers and waterfalls.

However, his job requires a lot more than choosing the colours and painting them on the wall. Mr Toppin must first decide how much paint is needed. This is calculated by knowing the total surface area of each wall, the total number of walls to be painted and the number of times each wall should be painted for the colour to show brightly. His creativity really comes out when Mr Toppin creates colours that have never been seen before. To do this, he mixes a little bit of one colour with one or two more colours until a totally new colour is made. In this way, his clients can paint their houses in colours made just for them.

Mr Toppin has to paint the walls of an office with 8 rooms, all the same size. The floor of a room has a width of 10 metres and a length of 15 metres. Each square metre of wall will require 15 ml of paint to complete the job. If the paint is sold in buckets containing 4 litres each, how many buckets will be needed to paint the entire office?

Unit 4.4 Time

In this unit you will:

- determine the time between events
- solve simple problems involving time.

Key words
12-hour clock
24-hour clock

Time was one of the first quantities to be measured. The hour glass and sundial were among the first devices used to measure time.

Hour glass

Some clocks and watches show time using a **12-hour clock**. On the clock alongside we cannot tell whether its 5 o'clock in the morning or 5 o'clock in the evening – these times look exactly the same on this clock.

Sundial

Modern digital watches often use **24-hour time**. In this system the times from midnight to 12 noon are represented using the numbers 0 to 12. The hours after noon to midnight are represented using the numbers 13 to 24. In this system 5 o'clock in the morning would be 05:00 hours and 5 o'clock in the evening would be 17:00 hours.

12-hour clock

Worked example

Express the time 20 minutes after 5 o'clock in the evening using:

a. the 12-hour system

b. the 24-hour system.

a. 5:20 p.m.

b. 17:20 h

Exercise

1 Write these times as they would appear on a 24-hour digital clock.

a) 8.25 a.m. **b)** 10.45 p.m. **c)** 2.18 a.m.

d) 3.05 a.m. **e)** 12.30 p.m. **f)** 12 midnight

2 Write these times using the 12-hour system and a.m. and p.m.

a) 09:35 h **b)** 16:20 h **c)** 20:10 h

d) 06:55 h **e)** 23:00 h **f)** 24:00 h

3 Calculate the difference between the first and second time in each case:

a) 3.15 a.m. and 7.20 p.m. on the same day

b) 6.30 a.m. and 2.45 p.m. on the same day

c) 04:45 hours on Monday and 13:00 hours on Tuesday

d) 11:50 hours on Sunday and 20:20 hours on Monday

Looking back

1. The heights of four boys are 1 m 10 cm, 1 m 20 cm, 1 m 5 cm and 1 m 35 cm. What is their total height in:

   a) metres and centimetres

   b) feet and inches?

2. A piece of string is 8 metres long. The string is cut into six pieces. Five of the pieces are each of length 1.3 metres. What is the length of the sixth piece, in inches?

3. Debra weighs 100 pounds and her sister weighs 42 kilograms. Determine the difference in their masses in kilograms.

4. A tour bus left the bus depot at 08:25 h and travelled for 6 hours and 48 minutes (including stops) before arriving at its destination in Mayaro. At what time did it arrive in Mayaro?

5. An athlete started to exercise at 25 minutes after midday. She swam for 50 minutes and then rested for 10 minutes. After this she ran for half an hour and then rode her bike for 35 minutes before stopping. At what time did she stop exercising?

6. A concert which lasted for 3 hours and 25 minutes ended at 18:10 h. At what time did it start?

7. A mountain climber climbed from the base of a mountain to a flag which was placed at the top of the mountain. The temperature at the base was 20 °C and dropped by 2 °C every 400 metres. If the flag was 2 800 metres from the base, what was the temperature at the top of the mountain?

8. The average temperature in a city in Canada for the first six months of the year is shown in the table below.

Month	January	February	March	April	May	June
Temperature (°C)	−6	−10	3	11	22	28

   a) Which month recorded the lowest temperature?

   b) What was the difference in temperature between the months of April and January?

   c) What was the average temperature for the six months?

9. Which is the longest measurement in each set?

a)	54.6 cm	570 mm	0.058 m
b)	98 mm	0.59 cm	0.0059 m
c)	600 mm	0.59 m	0.61 m
d)	1.2 km	12 000 m	1 200 000 mm
e)	8.4 m	849 cm	0.84 km

Chapter 5

Length and perimeter

In this chapter you will deal with standard units of lengths and the instruments we use to accurately measure lengths. You will have a chance to estimate and measure lengths of real and drawn objects. You will then extend your skills to measuring and calculating the perimeters of regular and irregular shapes. Later on, in Chapter 9, you will determine formulae for working with perimeter.

Key words
Metres
Centimetres
Millimetres
Kilometres

Unit 5.1 Measuring length

In this unit you will:

- identify standard units of linear measure
- identify and select instruments and appropriate units to measure length
- convert between SI units of linear measure
- accurately measure given lengths.

The Système International (SI) unit of length is the **metre**. All other units derive from the metre. (Remember a metre is just over three feet.)

At this level, metre lengths will be measured using a ruler, metre stick or measuring tape.

Shorter lengths can be measured in **centimetres** or **millimetres** using a ruler. Longer lengths are measured in **kilometres**. Mechanical devices are normally used to measure longer lengths.

The ruler below shows centimetres and millimetres. Can you see the relationship between these units? The line being measured is 45 millimetres, or 4.5 centimetres long.

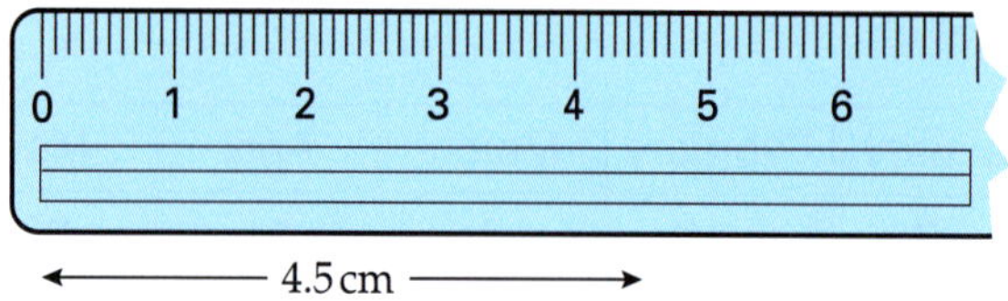

The table shows the relationship between units of length in the metric system.

Prefix	Meaning	Unit	Abbreviation
milli	$\frac{1}{1000}$ (one thousandth of a metre)	millimetre	mm
centi	$\frac{1}{100}$ (one hundredth of a metre)	centimetre	cm
deci	$\frac{1}{10}$ (one tenth of a metre)	decimetre	dm
kilo	1 000 (one thousand metres)	kilometre	km

Using the table we can see that it is possible to convert from one unit to another:

1 km = 1 000 m
1 m = 10 dm
1 dm = 10 cm
1 cm = 10 mm

Worked examples

1. Convert 36 mm to centimetres.
2. Convert 8 km to centimetres.
3. Convert 250 metres to millimetres.

1. 10 mm = 1 cm, so 36 mm = 36 ÷ 10 = 3.6 cm
2. 1 km = 1 000 m and 1 m = 100 cm, so 1 km = 1 000 × 100 = 100 000 cm.
   8 × 100 000 = 800 000 cm
3. 1 m = 100 cm and 1 cm = 10 mm, so 1 m = 100 × 10 = 1 000 mm
   250 × 1 000 = 250 000 mm

Note that to convert from a bigger unit to a smaller unit you multiply. To convert from a smaller unit to a bigger unit you divide.

Exercise

1 Write the following lengths in centimetres.

a) 23 mm **b)** 9.6 mm **c)** 3.4 dm **d)** 0.12 dm
e) 55 m **f)** 8.2 m **g)** 9 km **h)** 0.45 km

2 Write the following lengths in metres.

a) 23 800 mm **b)** 56 mm **c)** 650 cm **d)** 8 cm
e) 750 dm **f)** 5.4 km **g)** 9 km **h)** 0.33 km

3 Convert these lengths to kilometres.

a) 450 000 cm **b)** 8 000 cm **c)** 48 cm **d)** 65 dm
e) 720 m **f)** 2 500 m **g)** 0.53 m **h)** 0.004 m

4 What SI units would you use to measure:

a) the length of a shoe? **b)** the distance from home to school?
c) the thickness of this book? **d)** the length of a pencil?
e) the height of a mountain?

5 Work with a partner. Estimate the length of each line below and write down your estimates. Use a ruler and a piece of string to measure the lengths of the lines. Whose estimates were most accurate?

a) ______________________ **b)** ____________________________________

c) ____

d)
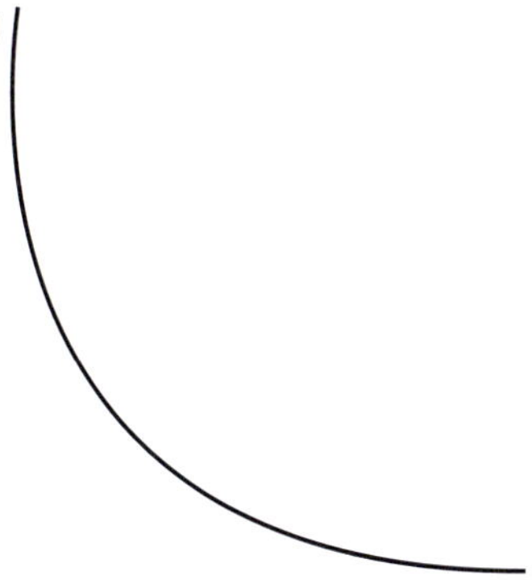

e)
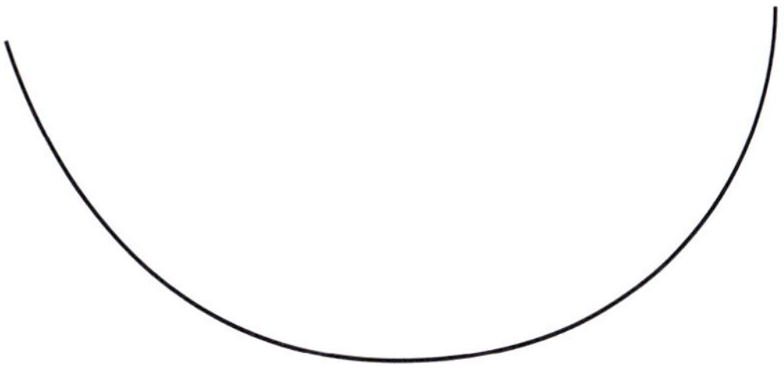

Unit 5.2 Perimeter

Key words
Perimeter
Regular

In this unit you will:

- determine the perimeter of regular and irregular shapes
- solve problems involving perimeter.

The **perimeter** of a polygon is the distance around all the edges of the shape. You can work out the perimeter by adding together the lengths of all the sides. For example:

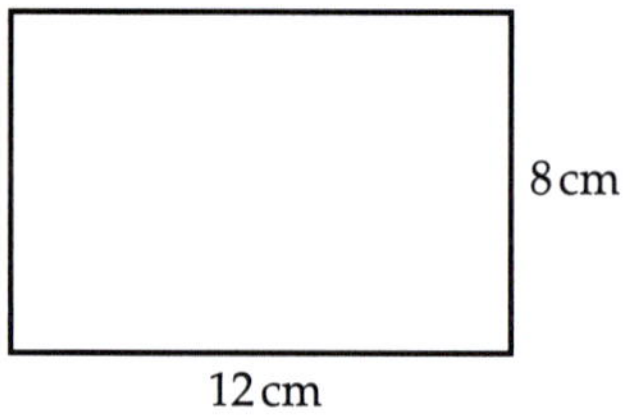

The perimeter of this shape is
12 cm + 8 cm + 12 cm + 8 cm = 40 cm

In a **regular** shape, all the sides are the same length. To work out the perimeter of a regular shape you can multiply the length of one side by the number of sides. For example:

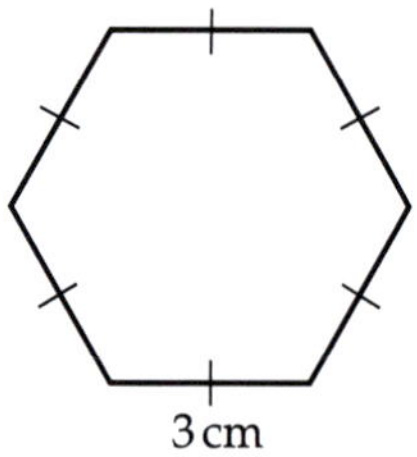

The perimeter of this shape is 3 cm × 6 = 18 cm.

If you know the perimeter of a regular shape, you can work out the length of each side by dividing by the number of sides.

Worked examples

1. Find the perimeter of shape ABCD.

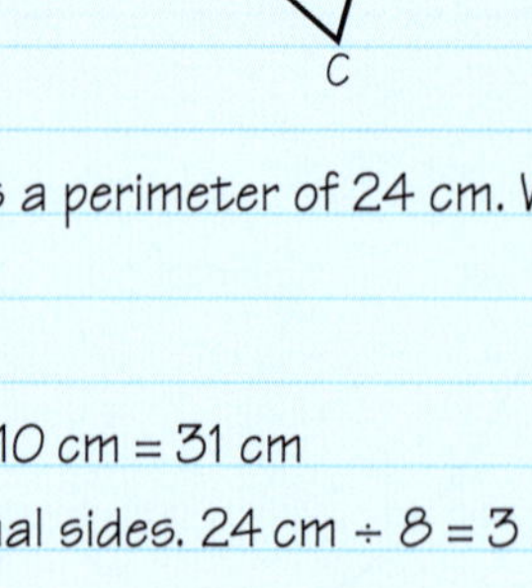

2. A regular octagon has a perimeter of 24 cm. What is the length of each side of the octagon?

1. 8 cm + 6 cm + 7 cm + 10 cm = 31 cm
2. An octagon has 8 equal sides. 24 cm ÷ 8 = 3 cm

Exercise

1. Calculate the perimeter of each shape.

a)

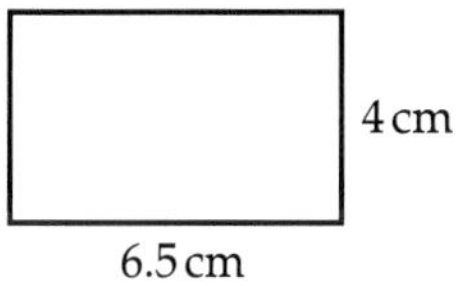

b)

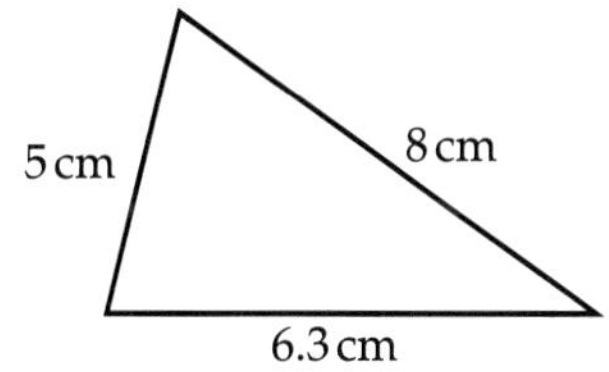

c)

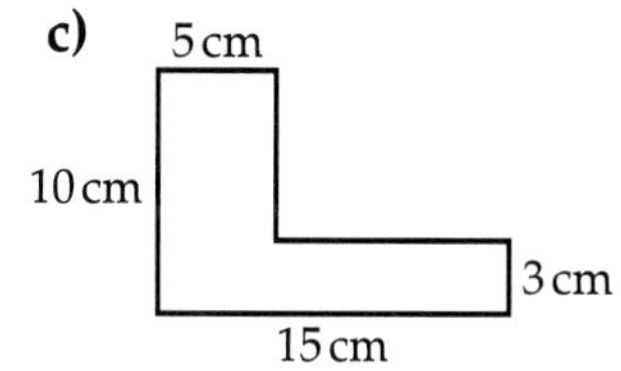

d)

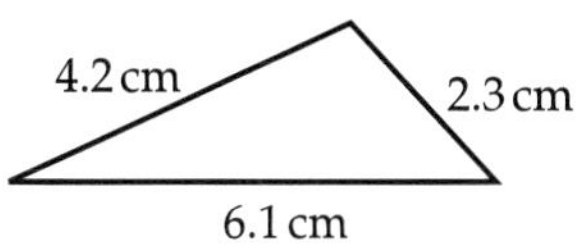

e)

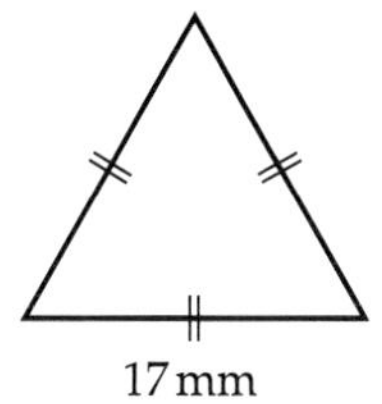

f)

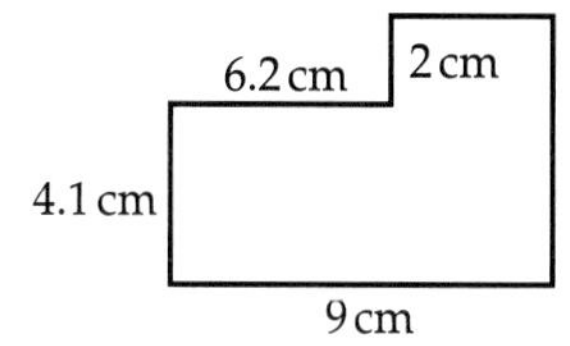

2. Calculate the perimeter of these shapes.
   a) a square of side 70 mm
   b) a rectangle of width 8 cm and length 15 cm
   c) an equilateral triangle with a side of 9 cm
   d) a regular pentagon with a side of 6.5 cm

3. A garden is in the shape of a rectangle. The length is 25 metres and the width is one quarter of the length. What is the perimeter of the garden?

4. Use the appropriate measuring instruments and units to measure and work out the perimeter of:
   a) the top of your desk
   b) the chalkboard
   c) the classroom ceiling
   d) a clock face
   e) a doughnut

5. A square has a perimeter of 100 m. What is the length of each side?

6. A rectangle of perimeter 100 cm has a length of 30 cm. What is its width?

7. An equilateral triangle has a perimeter of 240 cm. What are the lengths of the sides?

Investigation

How would you measure and work out the perimeter of a car tyre?

Find out what you call a doughnut or tyre shape in mathematics.

Looking back

1 Mrs Singh bought 1.65 metres of red ribbon, 3.2 metres of yellow ribbon and 86 centimetres of blue ribbon. How many metres of ribbon did she buy altogether?

2 An athlete ran a distance of 430 metres on Monday and 508 metres on Tuesday. How many metres should he run on Wednesday to run a total of 1.5 kilometres for the three days?

3 A textbook is 32 millimetres thick. If 45 of these books are stacked on top of each other, what is the total height, in metres?

4 A piece of string is 7.2 metres long. How many pieces each of length 24 centimetres may be cut from the string?

5 The perimeter of a rectangle is 24.8 cm. If one side is 5.6 cm, what is the length of the other side, in millimetres?

6 A hockey field is 90 metres long and 45 metres wide. The field is divided into three equal sections by marking three points along the length AB. What is the perimeter of each section of the field?

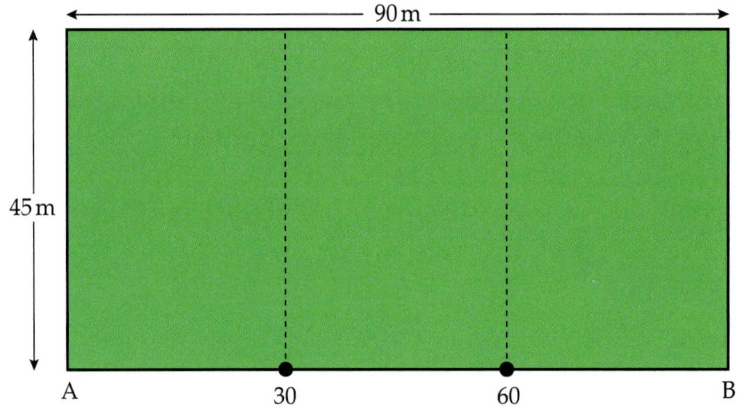

7 A tri-athlete trains every Saturday by running 2 km and cycling for 5 km. If he swims half the distance that he runs, what total distance does he cover, in kilometres?

8 The table below shows the distance in kilometres a helicopter flies between five towns.

Town	Port-of-Spain	Arima	San Fernando	Scarborough	Point Fortin
Port-of-Spain	0	28	43	96	55
Arima	28	0	45	80	68
San Fernando	43	45	0	136	28
Scarborough	96	80	136	0	152
Point Fortin	55	68	28	152	0

a) What is the distance from Arima to Scarborough?

b) Which town is the closest to San Fernando?

c) What is the total distance from Port-of-Spain to Arima via Point Fortin?

Chapter 6

Key words
Area
Surface
Square units

More about measuring

You have already measured different lengths. In this chapter, you will combine what you know about measuring lengths and about properties of shapes to find the area of flat shapes. The formulae for finding the area of some shapes will be explained and applied.

Unit 6.1 Area

In this unit you will:

- know and use units of area
- estimate areas of shapes by counting squares.

Area is the total space contained within the edges of a shape.
Look at the photographs carefully. In Photograph A, the area of the window is the **surface** covered by glass. In Photograph B, the area of the kitchen floor is the surface covered by tiles.

Photograph A

Photograph B

Area is measured in **square units**. The diagrams below show a square millimetre (mm^2) and a square centimetre (cm^2):

1 mm^2 1 cm^2

We can also measure in square metres (m^2) and square kilometres (km^2).

The easiest way of finding the area of a shape is to count the square units. For example:

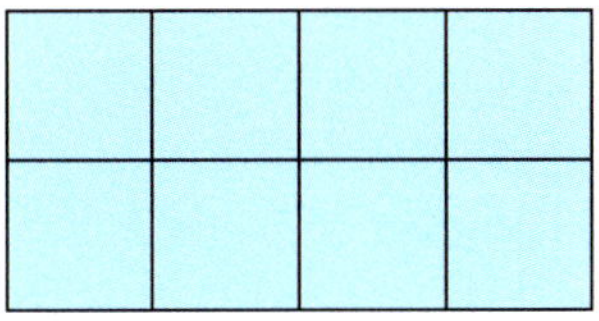

This rectangle has an area of 8 blocks. Each block is 1 cm^2, so the area is 8 cm^2.

1 square metre (m^2) is equal to the area of a square with sides of length 1 metre. The top of a school desk is likely to be about 1 m^2. Four tiles on the floor of the kitchen in the photograph on page 55 have an area of 1 m^2.

1 square kilometre (km^2) is equal to the area of a square with sides of length 1 kilometre. This unit is used to measure larger areas such as land and city blocks and countries. For example, the area of Trinidad and Tobago is approximately 5 130 km^2.

Worked examples

Each square represents 1 cm^2.

1. Determine the area of shape **a**.
2. Estimate the areas of shapes **b** and **c**.

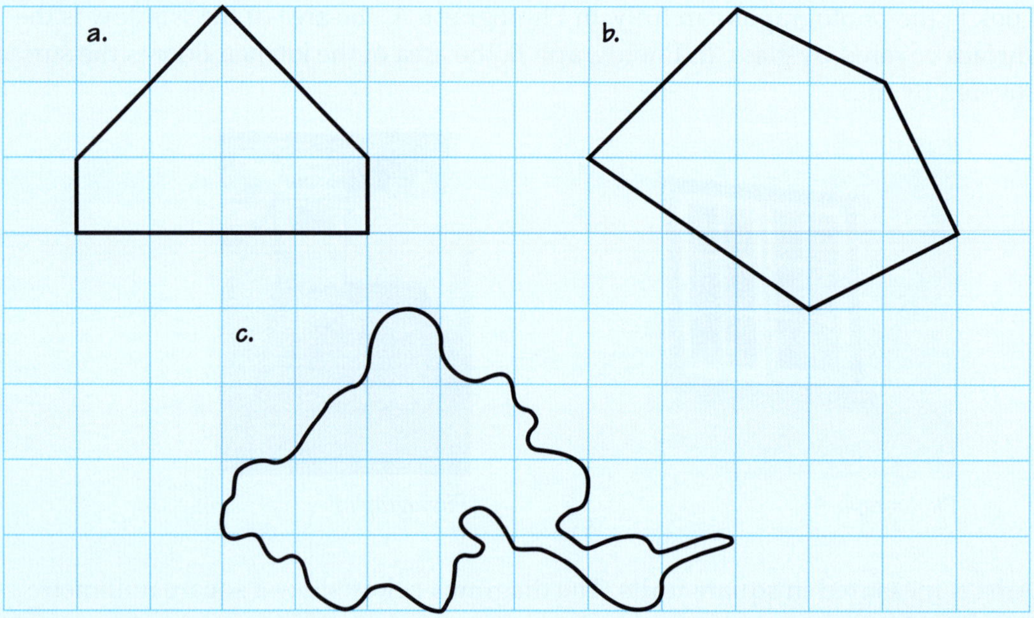

1. a. 6 whole squares = 6 cm^2
   4 half-squares = 2 cm^2
   Total area = 8 cm^2

2. b. 5 whole squares = 5 cm^2
   7 squares with half or more shaded ≈ 7 cm^2
   Total area ≈ 12 cm^2

   c. 4 whole squares = 4 cm^2
   8 squares with half or more shaded ≈ 8 cm^2
   Total area ≈ 12 cm^2

Count whole squares.
Add half-squares to make whole squares.
Count any part that is half or more as a whole square.
Ignore parts that are less than half a square.

Exercise

1. State which units of area (mm^2, cm^2, m^2 or km^2) would be most suitable for measuring the area of:

   a) your island

   b) a tile on the classroom floor

   c) this book's cover

   d) a postage stamp

   e) a sheet of writing paper

   f) the surface of the Caribbean Sea

   g) your little toe nail

   h) the Antigua cricket grounds

   i) a small watch face

   j) a pizza

2. The following shapes have been drawn on squared paper. Four small blocks are equal to 1 cm^2. Find the area of each shape in cm^2. Ignore all shaded sections.

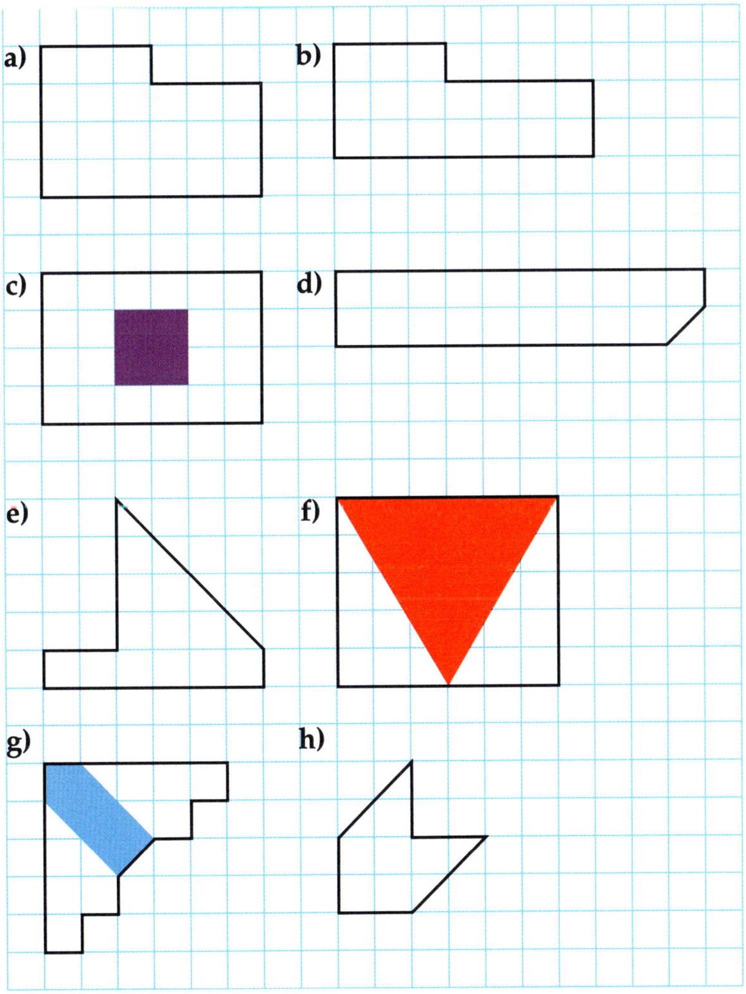

Unit 6.2 Formulae for area

Key words
Length
Breadth
Base
Height

In this unit you will:

- learn the formulae for calculating the area of rectangles and triangles
- apply formulae to find the area of basic and compound shapes
- solve word problems related to area.

The area of a rectangle may be calculated by multiplying the **length** by the **breadth**. Breadth is another name for width. This is the formula for area of a rectangle:

Area = length × breadth OR $l \times b$

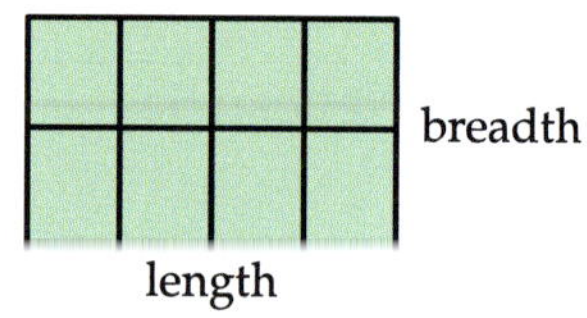

Since a square is a rectangle with all its sides equal, the area of a square is length × length. This is the formula for area of a square:

Area = length × length OR l^2

A right-angled triangle can be made by dividing a rectangle in half. This means its area is half of the area of a rectangle. In fact, all triangles are half of a rectangle. We use this knowledge to derive a formula for area of a triangle:

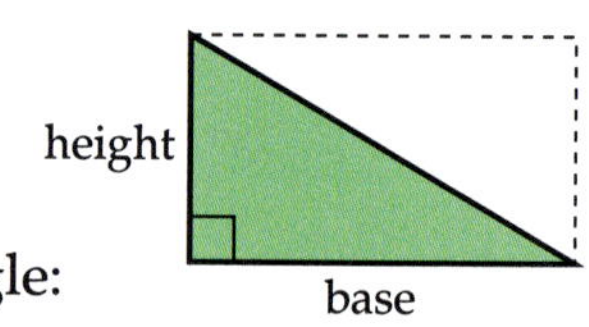

$$\text{Area} = \frac{\text{length} \times \text{breadth}}{2}$$

For triangles this is normally written as: Area = $\frac{1}{2}$ **base** × **height**.

Worked examples

Find the area of these shapes:

1.

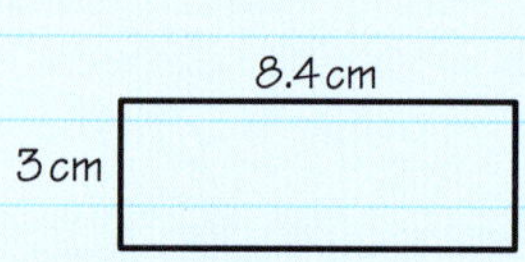

2.

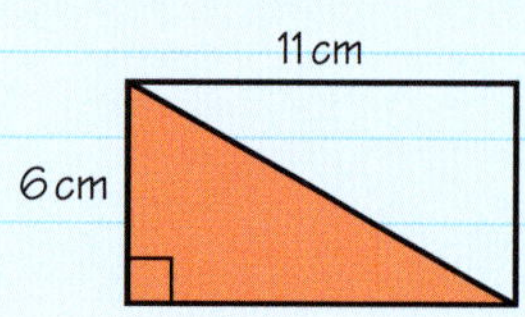

3.

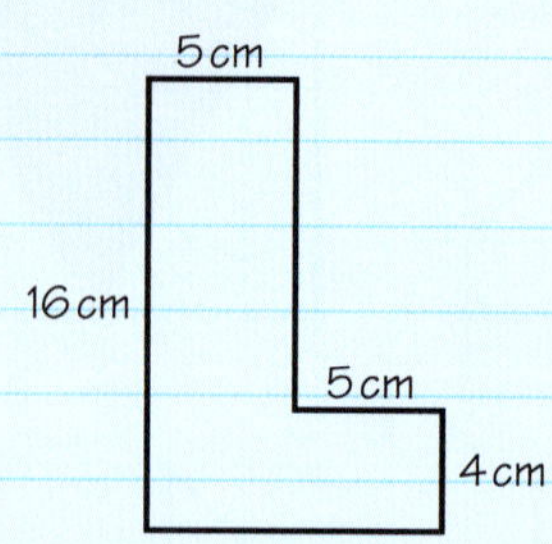

1. Area of a rectangle $= L \times B$

$= 8.4\text{ cm} \times 3\text{ cm}$

$= 25.2\text{ cm}^2$

2. Area of the rectangle is $L \times B = 6\text{ cm} \times 11\text{ cm} = 66\text{ cm}^2$

Area of triangle is $\frac{L \times B}{2} = \frac{66\text{ cm}^2}{2} = 33\text{ cm}^2$

3. Divide the shape into two rectangles:

$A = 5\text{ cm} \times 16\text{ cm} + 5\text{ cm} \times 4\text{ cm}$

$= 80\text{ cm}^2 + 20\text{ cm}^2$

$= 100\text{ cm}^2$

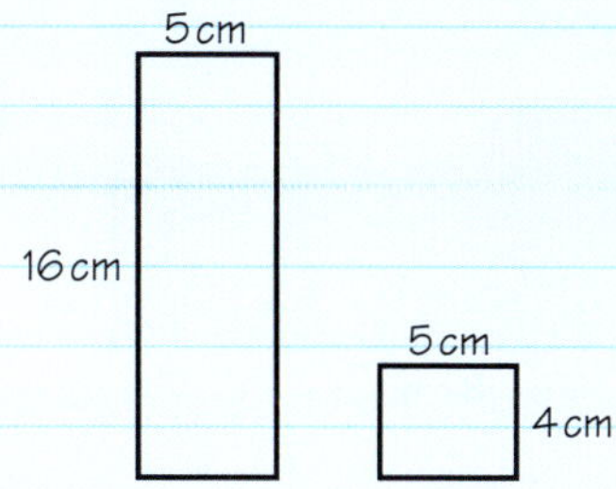

Exercise

1. Apply formulae to find the area of each of these shapes:

a)
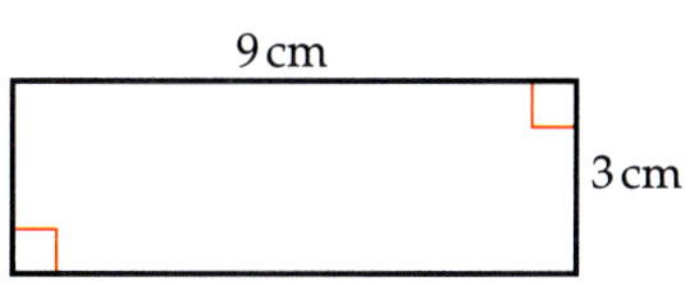

b)
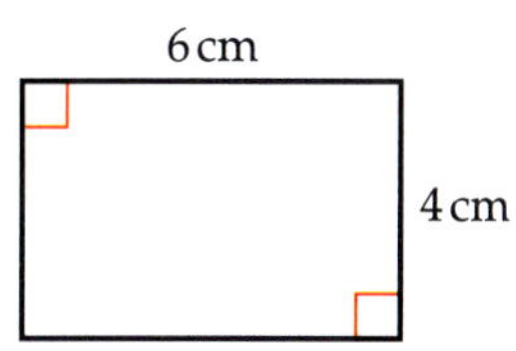

c)
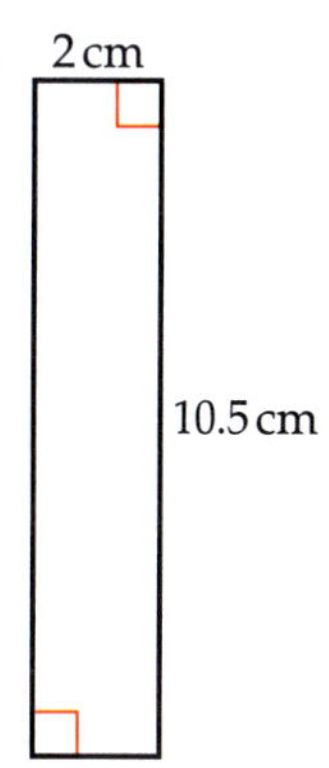

d)
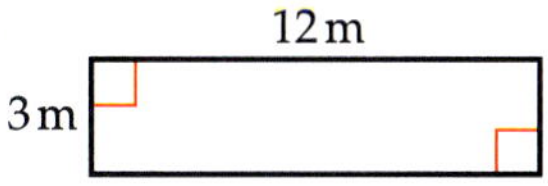

e)
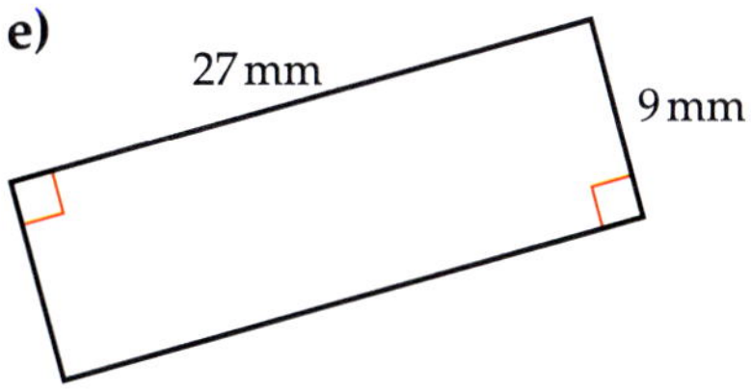

f)
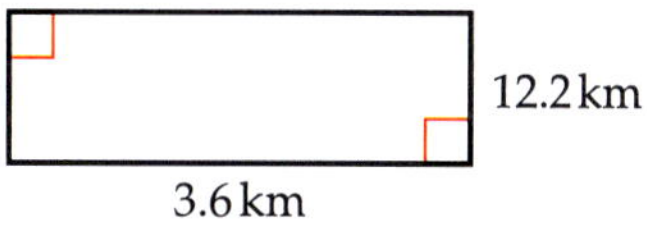

g)
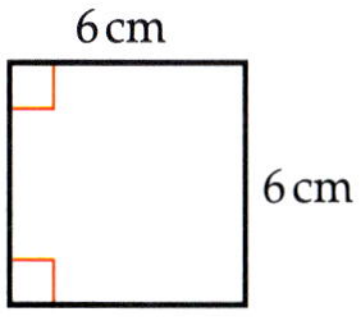

h)
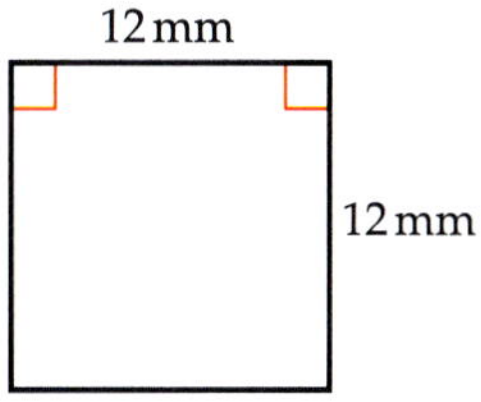

i)
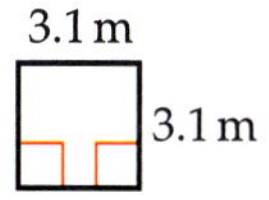

j)
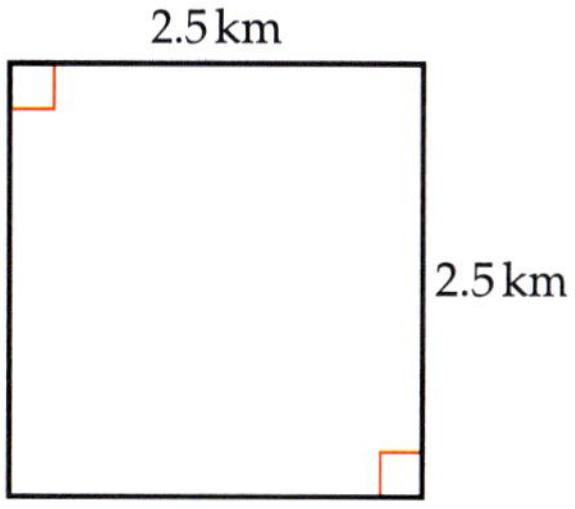

2 Find the area of these triangles by considering them to be half a rectangle:

a)

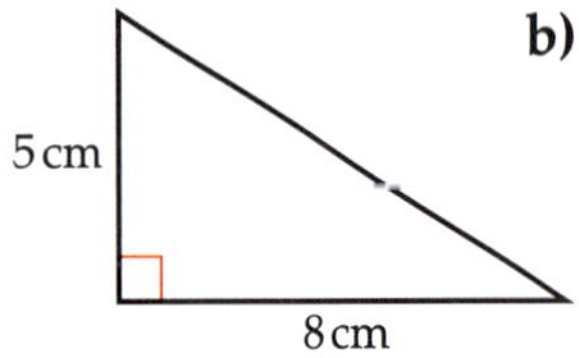

b)

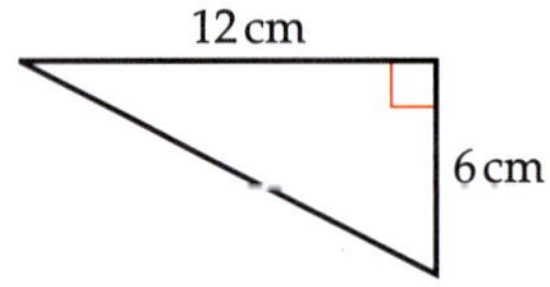

c)

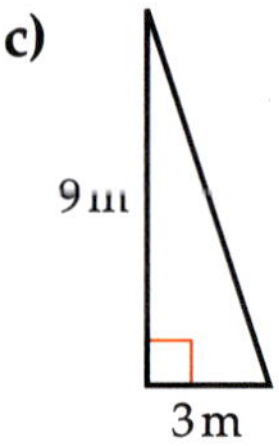

d)

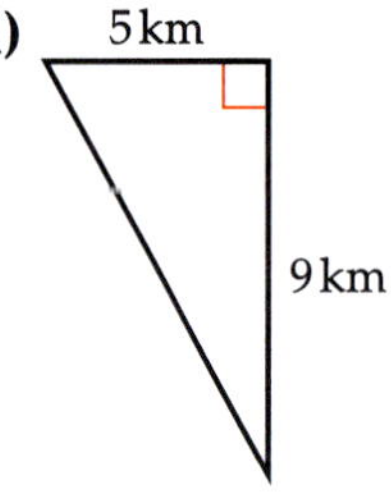

3 Determine the area of the shaded region in each shape. Divide the shapes into smaller parts to help you.

a)

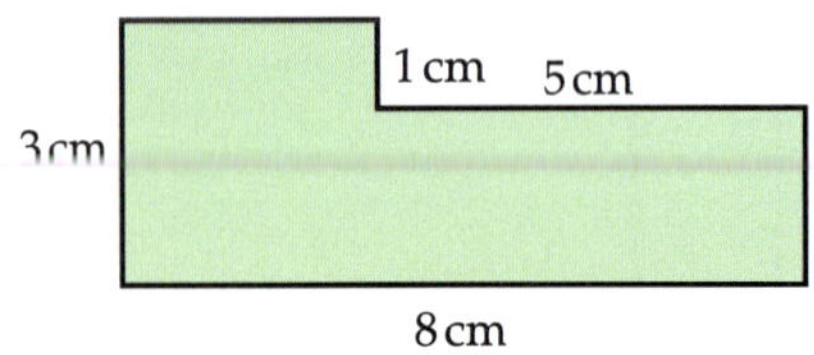

b)

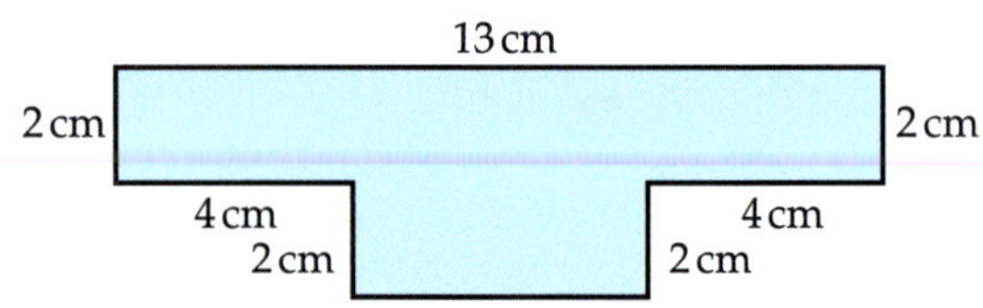

c)

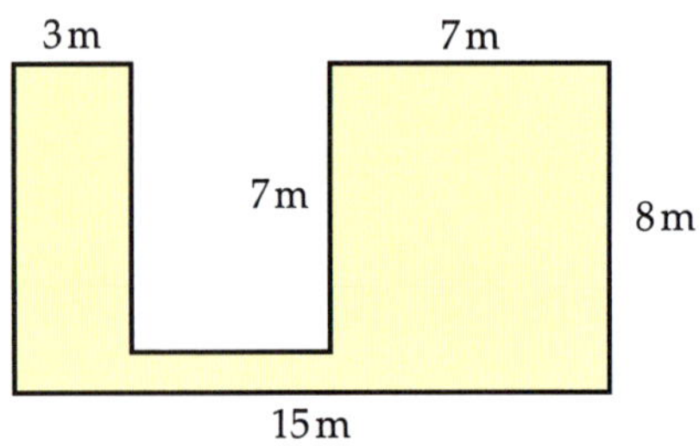

d)

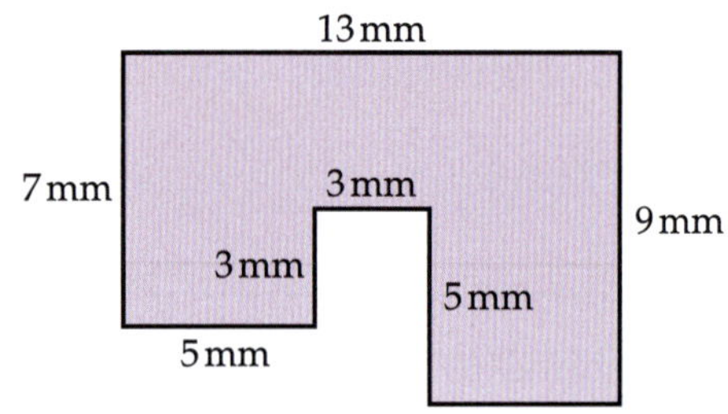

e)

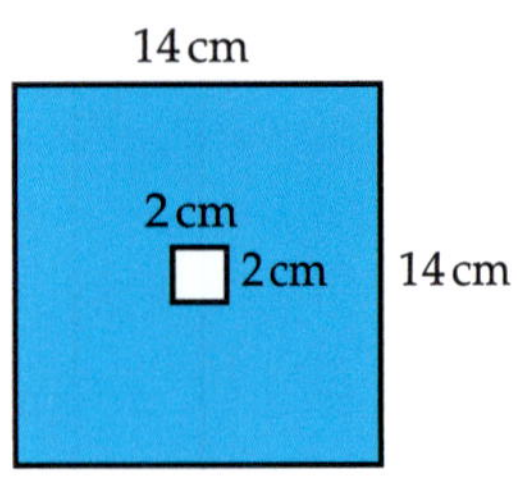

f)

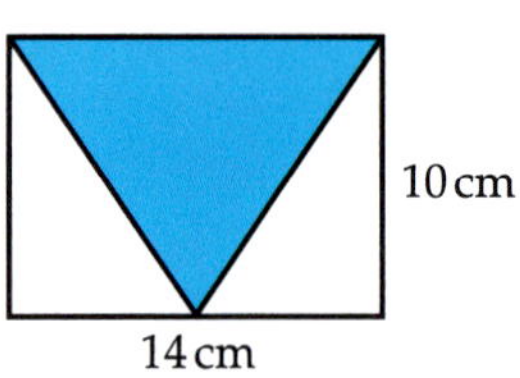

4 The length of a rectangle is 6 cm and its area is 22.8 cm^2. What is the width of the rectangle?

5 The area of a square is 81 cm^2. What is the length of each side of the square?

6 The perimeter of a rectangle is 46 cm. If the length of the rectangle is 8 cm, calculate its area.

7 A rectangle has a base of 30 mm and a height of 5.5 cm. What is the area of the rectangle in cm^2?

8 A triangle has a base of 12 cm and a height of 6.5 cm. Calculate the area of the triangle.

Investigation

For each of the triangles below:

a) Measure the length of the base.

b) Measure the height of the triangle.

c) Determine the area of the triangle in square units by counting the squares.

d) Calculate the area of the rectangle around the triangle.

Triangle A has been done for you as an example:

a) Base is 8 units.

b) Height is 5 units.

c) Area of triangle is 20 square units.

d) The area of the rectangle is 40 square units.

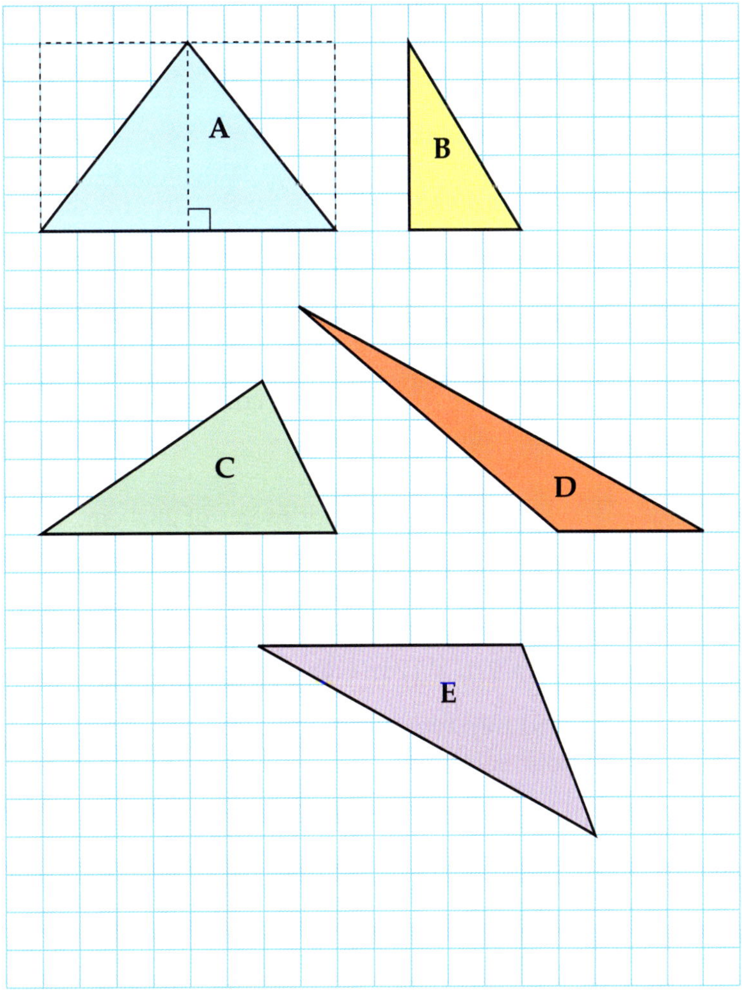

What do you notice about the area of each triangle and the rectangle around it?

Looking back

1 Find the area of each shape.

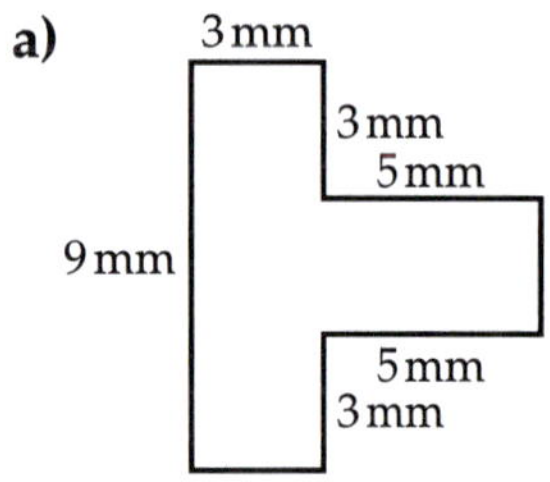

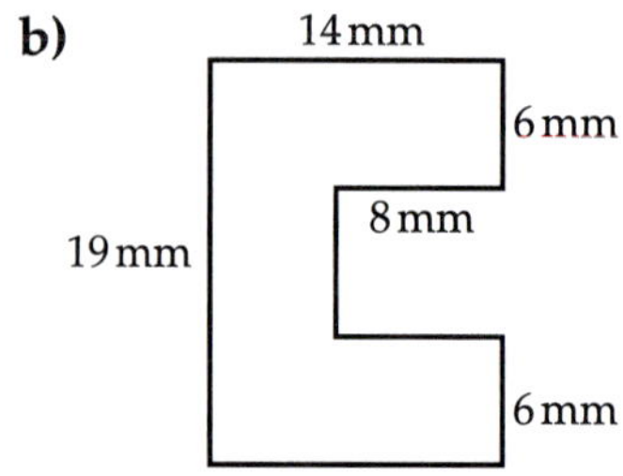

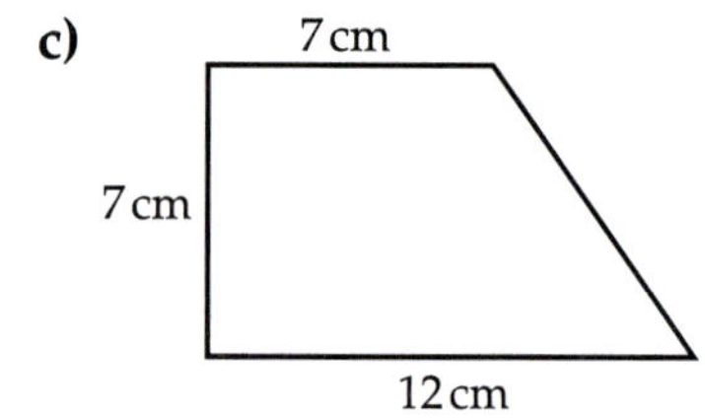

2 Find the shaded area.

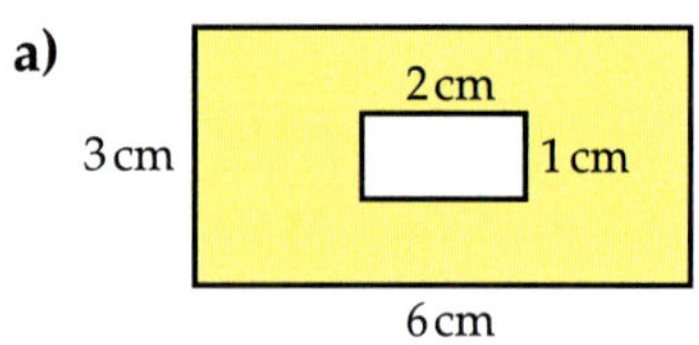

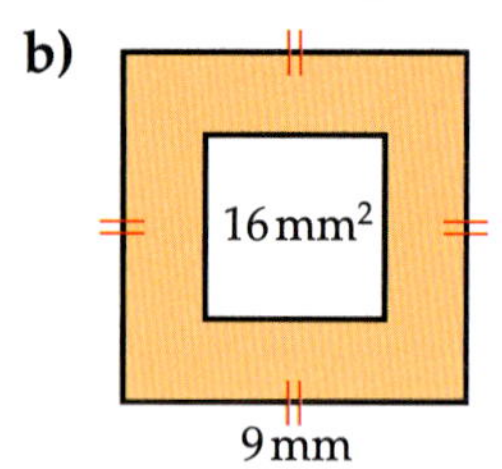

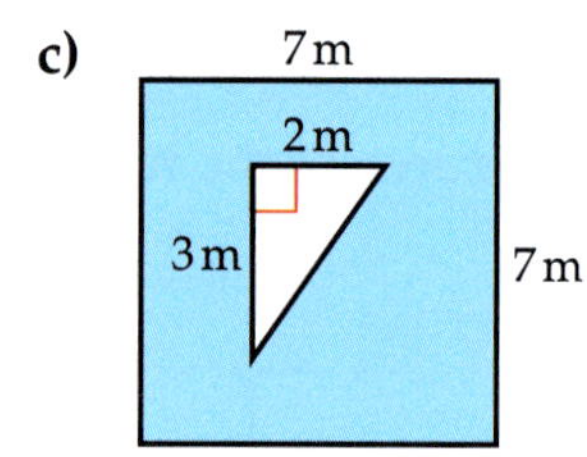

3 A picture frame measures 32 cm × 48 cm on the outside. The picture inside the frame measures 28 cm × 24 cm. What is the area of the border around the picture?

4 A floor has a length of 20 metres and a width of 15 metres. How many square tiles each of length 50 cm are needed to cover the floor completely?

5 The height of a triangle is 77 mm and its base is 8 cm. Calculate the area of the triangle in square centimetres.

6 Determine the area of a square which has a perimeter of 28.4 cm.

7 After a bad storm in Tobago in 2004, the wall of a building was left with a rectangular hole measuring 6 m × 8 m.
One piece of wood measuring 4 m × 12 m was delivered to patch the hole.

How could you cut the wood into two pieces to patch the hole?
Draw diagrams to show your solution.

Project

You will need a ruler with cm and mm divisions to complete this project.

1. Estimate the length of each horizontal line below and record your estimates in the first column of a table like the one below the lines.

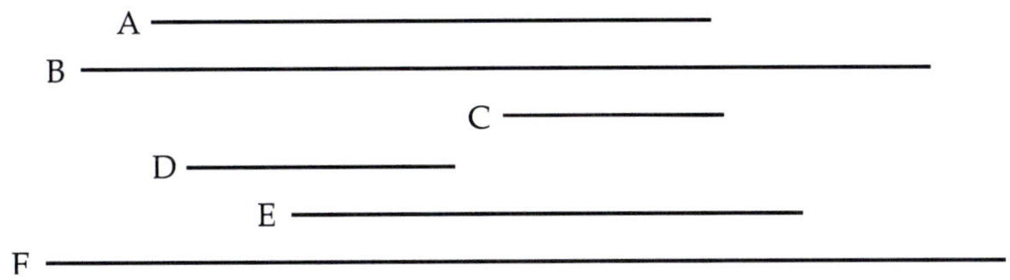

Line	Estimate of length	Measured length	Difference between estimate and measured length
A			
B			
C			
D			
E			
F			

2. Now measure each line in millimetres and complete the second blank column of your table.

3. For each line, calculate the difference between your length estimate and the measured length and complete the last column of the table.

4. Add the values in the last column. This is your total error.

5. Draw up another table like the one above and follow the steps again, but this time use the vertical lines opposite:

6. For which set of lines did you record the largest total error? Can you suggest why?

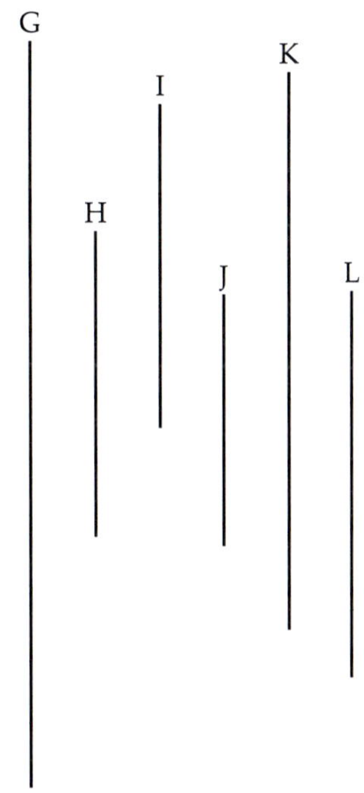

Chapter 7

Using algebra

This chapter introduces some of the concepts that you will need as you develop your skills in algebra. You will learn about variables, coefficients, terms, expressions and equations so that you can understand and use these to solve problems.

Key words
Algebra
Variable
Expression
Term
Coefficient
Constant

Unit 7.1 Using symbols for numbers

In this unit you will:

- use letters to represent numbers
- identify variables, coefficients, terms and expressions.

In maths we often use a letter to stand in for a number we don't know. This is called **algebra**.

If you chose a number and added 5, you could represent this by writing $x + 5$. It doesn't matter what letter or symbol you use to represent the unknown quantity.

The value of x can change, so we call x a **variable**.

We call $x + 5$ an expression. An **expression** is a group of terms. In algebra **terms** are separated by + or − signs. They are not separated by ÷ and × signs.

$2y^2 + 6xy - 7$	This expression has three terms: $2y^2$, $6xy$ and -7.
$2 \times 5a$	This expression has one term. It is the same as writing $10a$. **Terms are not separated by × signs.**
$7y^2 \div 6x$	This expression has one term. **Terms are not separated by ÷ signs.**
$5x^3 + 7x^2 \div 3y - 2xy + y$	This expression has four terms: $5x^3$, $7x^2 \div 3y$, $-2xy$ and y.

We call the numeral in front of the variable the **coefficient**. For example, in the term $80a$, the coefficient is 80.

In the term $-2xy$, the coefficient is -2. The sign remains with the coefficient.

In the term xy the coefficient is 1. We do not normally write the 1 because it doesn't change the value.

The value of ordinary numerals in an expression does not change, so we call such numerals the **constant** term. For example in the expression $2x + 7$, the constant is 7.

Exercise

1. Write the coefficients in the following terms.
   a) $3y$ b) $-2x$ c) $6a$ d) $-4b$
   e) $36p$ f) $-p$ g) $\frac{1}{2}m$ h) $0.5z$

2. In the expression $5x - 12$:
   a) What is the coefficient of x? b) What is the constant term?

Making your own expressions

At a family reunion Mr Mighty asked his great-grandchildren to tell him the age of each person in the room. Judith replied, 'I don't know their ages but I know that when I was born each person was $n - 10$ years old.' James said, 'I know that in two years' time we will all be $n + 2$ years old and I will be as old as Judith is now'.

Notice that the children used the letter n for the ages they did not know.

Ages of some guests at the reunion party

Guest	Age when Judith was born (years)	Age at time of party (years)	Age in two years' time (years)
Judith	0	10	12
Mr Mighty	80	90	92
James (Judith's brother)	Not yet born		10
Mr Mighty's wife			87
Judith's mother	26	36	
Judith's father	28		
Any guest	$p - 10$	p	$p + 2$

The unknown p is given for the present age of each person at the party.

We can use algebraic expressions to represent mathematical problems.

Worked example

Write the following as expressions using algebra:

a. I choose a number and multiply it by 7.

b. Janet has some marbles in her pocket. She removes 5 of them. How many marbles are now in her pocket?

c. James has some sweets. If he shares them equally among five people, how many sweets does each person have?

a. $7x$ Let the number be x.

b. $m - 5$ Let the original number of marbles be m.

c. $\frac{s}{5}$ Let the number of sweets be s. $\frac{s}{5}$ means the same as $s \div 5$.

Exercise

1. Copy and complete the table of ages above.

2. Write the following expressions using x to stand in for the number I choose.

   a) I choose a number and subtract 5. **b)** I choose a number and triple it.

   c) I choose a number and add 32. **d)** I choose a number and divide it by 7.

Unit 7.2 Making equations

Key words
Equation

In this unit you will:

- write simple equations using letters to stand in for unknown quantities.

Tom has a certain amount of money. We'll call it y. His sister has twice as much money plus \$5 more. Together they have \$140.

This story problem can be written as:

$y + (2y + 5) = 140$

This is called an **equation** because it includes an equal sign. An equation is a complete number sentence with an equal sign. Algebraic equations contain at least one variable.

Worked examples

1. Two whole numbers added together make 15. Write an equation to show this using the letters a and b to represent the whole numbers.
2. I choose a number, multiply it by 3 and get 150. Write an equation to help you find the number I chose.

1. $a + b = 15$
2. $3n = 150$.

We write $3 \times n$ as $3n$.

Exercise

1 Say whether each of these is an expression or an equation.

a) $h + 45 - 21p$

b) $l + b = \frac{1}{2}p$

c) $p + q + 5r - 30$

d) $I = \dfrac{PTR}{100}$

e) $2a + 3b = 56$

f) $xy - 3x + 4y - 8$

g) $28 = 2 + 4c$

h) $61 + F = G$

i) $Aa + b + c - 3Aa + 2b$

j) $y = mx + c$

k) $4h = 56$

l) $\dfrac{5a + b}{3a - b} + 24b + 6$

2 For each of the following, use any variable to write equations or expressions. Remember to state the unknown that the variable is replacing.

a) I think of a number, multiply it by 20 and subtract 18.

b) Jojo has three times his brother's marbles; together they have 48 marbles.

c) Mr King takes half the time his son takes to mow the lawn. One month Mr King mowed once and his son mowed twice. How many hours are spent on mowing the lawn that month?

d) The sum of three consecutive numbers is 45. (x, $x + 1$ and $x + 2$ are consecutive numbers.)

e) An apple costs \$2.00 more than a mango. The total cost of three apples and a mango is \$50.00.

f) Write the total age for Sue, Beth and Mark if Sue is 3 years older than Beth and Beth is 5 years younger than Mark.

3 Write story problems for the following expression or equation. The unknown is given.

a)	$5h - 16 = h$	h is a number
b)	$s + 19$	s is the year of a storm
c)	$150 = 2\bigstar + 3\blacktriangle$	$\bigstar$ and $\blacktriangle$ are lunch money for two boys
d)	$8m - 5 = 30$	m is a number of cherries in each bag for sale
e)	$90 - x$	x is the value of an angle

Looking back

1 How many terms are there in each expression?

a) $2x$ **b)** $2x + ab - 5$ **c)** $x^3y - 4y$

d) $15b + 4$ **e)** $x^2 - y^2 + r^2 + x$ **f)** $8(a)(b)$

g) $5x^2$ **h)** $-x + 3y$ **i)** $x^2y^2 + 3y$

j) $5x^3 + x^2 - 3$ **k)** $6x^2 + x \times y$ **l)** $3px + 2x - p$

m) $2y^2r + y^2 \div 2$ **n)** $\frac{5}{2}x \times 3$ **o)** $4y^3 \times 2y + y^3$

2 Georgina, Raashad and Emily are all given the same amount of money. Use the letter b to stand for the number of cents they are given.

a) Georgina spends 45c of her money. Write an expression to show how many cents she now has.

b) Raashad shares his money equally with his brother. Write an expression to show how many cents he has left.

c) Emily is given \$1 more. Write an expression to show how many cents she has now.

Remember \$1 = 100 cents.

3 Write the following using algebraic expressions:

a) Ravi is h m tall. What is his father's height if he is 60 cm taller than Ravi?

b) During carnival, a moko jumbie took steps that were 1.8 m long. How far did he travel in s steps?

c) What is a child's mass if she is half her mother's mass of x kg?

d) What is the mass of three mangoes that weigh x g, y g and z g respectively?

Investigation

Work in pairs. Each of you should make up five story problems.

Exchange these with your partner and write equations which you will use to solve each problem.

Chapter 8

Simple algebra

This chapter will help you to apply the four basic operations to algebraic expressions. You will learn more about terms and how to simplify and expand them and will then simplify expressions by applying the basic rules of operations.

Key words
- Like term
- Unlike term
- Power
- Index/indices
- Expand
- Simplify

Unit 8.1 Like and unlike terms

In this unit you will:

- learn to differentiate between like and unlike terms
- use the four basic operations to simplify expressions.

Like terms have exactly the same variables. In other words, the letters and **powers** of unknowns in the terms must match exactly. The following are sets of like terms:

- $2a, 6a, 20a, a$ — a is the same as $1a$.
- $4ab, 3ba$ — ab means $a \times b$, so ab and ba are like.
- $2p^2, 15p^2$ — p^2 is read 'p squared'. It means $p \times p$. The ^{2} is an **index**. The plural of index is **indices**.

$8a$, $2a^3b$ and $3a^2$ are **unlike terms**. The letters or powers in the terms are not identical.

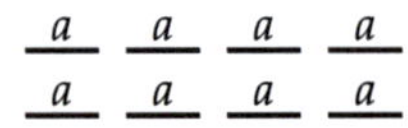

Eight lines of length 'a'

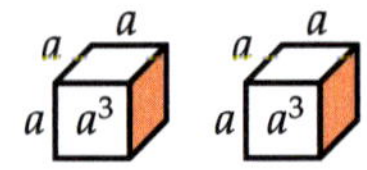

Two cubes of volume a^3

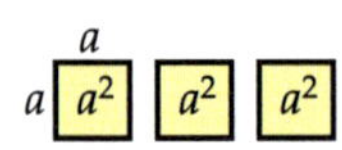

Three squares of area a^2

When we **expand** a term, we write it out in full using operation signs. For example:

$6a = 6 \times a$ or $a + a + a + a + a + a$

$4pq = 4 \times p \times q$

$s^3 = s \times s \times s$ — We say 's cubed'. The index tells you how many times a variable is multiplied by itself.

$3t^6 = 3 \times t \times t \times t \times t \times t \times t$ — We say 'three t to the sixth power'.

To **simplify** an expression, we write it in the shortest possible form and collect like terms. For example:

$3y + 12y - 18y + 5y = 2y$ — Add or subtract the coefficients of like terms.

$v \times v \times v \times v \times v \times u \times u = v^5 \times u^2 = v^5u^2$ — We leave out the multiplication sign in algebraic terms.

We can represent terms using geometric diagrams. For example, $6p$ may be a rectangle with a length of 6 units and a breadth of p. It could also be a line divided into six equal parts that are p units long.

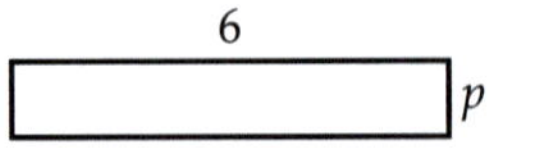

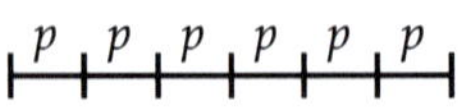

Worked examples

1. The diagram shows you the cost of various items and the table shows you what three different children bought.

Child	Purchases
Jack	3 pies, 1 cup-cake, 1 bottle of milk
Jane	1 pie, 1 bottle of orange juice, 2 cup-cakes
June	2 pies, 2 cup-cakes, 1 bottle of milk

   a. Write expressions to show the cost of the items bought by each child.

   b. Write one expression to show the total cost of all the purchases.

2. Simplify:

   a. $15a + 12b + 6a - 10a + 4b - 5b$ b. $7d^2 + 2d + 8d^3 - 3d^2 - 5d^3 + d$

   c. $7xy - 9x + 2y - 5yx + 14x$

1. a. Jack spent $(3y + r + t)$ dollars; Jane spent $(y + p + 2r)$ dollars and June spent $(2y + 2r + t)$ dollars.

   b. The total spent was: $(3y + r + t) + (y + p + 2r) + (2y + 2r + t)$ dollars

   $= 3y + y + 2y + r + 2r + 2r + t + t + p$

   $= 6y + 5r + 2t + p$ Collect like terms and add them.

2. a. $15a + 6a - 10a + 12b + 4b - 5b$ Keep the + or − sign with the term that follows it.

   $= 11a + 11b$

   b. $8d^3 - 5d^3 + 7d^2 - 3d^2 + 2d + d$ Collect terms with the same indices.

   $= 3d^3 + 4d^2 + 3d$

   c. $7xy - 5yx + 14x - 9x + 2y$ Remember that $xy = yx$.

   $= 2xy + 5x + 2y$

   Use the alphabetical arrangement in the final answer, so write xy rather than yx.

Exercise

1. Expand the following:

   a) $5p$ b) $k8$ c) $8b$ d) $g10 + h^4$ e) $9y + 4z$

2. Simplify:

   a) $a + a + a + p + p + p + a + a + a$ b) $g + g + h + h + h + h + h + g + g$

   c) $m \times m \times m \times m \times m \times m \times m \times n \times n \times n$

3. Simplify the following:

   a) $15m + 26m - 30\,m$

   b) $4p + 19q + 28p - 16q$

   c) $25h + 5r + 18hr - 6r - 12hr + 11r$

   d) $2h^2 - 40p + 50p^2 + 75p + 15h^2 - 51p^2$

   e) $6a^2 + 12ab + 14b^2 - 3a^2 - 11ab - 4b^2$

   f) $k^3 + 2k^2 + 19k^3 + 7k^3 - 2k^2 - 19k^2 + 20k^2$

   g) $efg + 4fg - 2ef + 16efg - 12fg + 3ef - 8fg$

   h) $55p + 24 + 2pr - 11p - 20$

Unit 8.2 Substituting in terms and expressions

Key words
Substituting
Value
Term
Expression

In this unit you will:

- calculate the numerical values of algebraic terms and expressions by substituting values for unknowns.

In the previous unit you expressed what each child spent as algebraic expressions with unknowns (variables). This table shows the actual price of each item.

Cost of item	
1 pie (y)	\$20.00
1 bottle milk (t)	\$10.00
1 bottle orange (p)	\$15.00
1 cup-cake (r)	\$5.00

This allows us to work out the actual amounts by **substituting** the real values for the unknowns.
Let us use Jack as an example: he spent $(3y + r + t)$ dollars.

This means he spent: $(3 \times \$20) + \$5 + \$10 = \$75$.

The **value** of a **term** or an **expression** is a number. For example, the value of $6a$ when $a = 7$ is $6 \times 7 = 42$.

Substitute means to put values in place of unknowns. For example, when $p = 15$, $r = 5$, $t = 10$ and $y = 20$ we can substitute these values in the expression $6y + 5r + 2t + p$ to find the total spent:

$$6(20) + 5(5) + 2(10) + 15 = 180$$

When you substitute, replace the values in each term and then simplify the numerical expression.

Worked examples

1. Find the value of $6pqr - 5qr$ when $p = 4$, $q = 2$ and $r = 5$
2. Find the value of $2d^3 - 4d^2 - 5d$ when $d = 3$
3. When $s = 1$, $f = 8$, find the value of $18sf + 6fs + 10s - 5f + 3s$

1. $6pqr - 5qr = 6 \times 4 \times 2 \times 5 - 5 \times 2 \times 5 = 240 - 50 = 190$
2. $2d^3 + 4d^2 - 5d = 2 \times 3 \times 3 \times 3 - 4 \times 3 \times 3 - 5 \times 3 = 54 - 36 - 15 = 3$
3. Sometimes it is easier to simplify first.

   $18sf + 6fs + 10s - 5f + 3s = 18sf + 6fs + 10s + 3s - 5f = 24fs + 13s - 5f$ Simplifying first.

   $= 24 \times 8 \times 1 + 13 \times 1 - 5 \times 8 = 192 + 13 - 40 = 165$ Substituting.

Exercise

1 Find the value of the following expressions. The values of the unknowns are in the square brackets.

a) $10f - 4f + 3f$ [$f = 5$]
b) $42g - h - 17j$ [$g = 7, h = 8, j = 9$]
c) $w^3 + w^2 + 2w$ [$w = 4$]
d) $2pq - 6p - q + 8p$ [$p = 9, q = 6$]
e) $3x^2y - 4xy^2$ [$x = 8, y = 7$]

Key words
Brackets
Expand
Distributive rule
Vinculum

Unit 8.3 Using brackets

In this unit you will:

- use brackets to simplify and expand algebraic expressions.

Brackets show that there is a close relationship; therefore what is in a bracket must be calculated first.

Brackets help us to work with the order of operations. For example, the brackets in the expression $5(a + b)$ can be expanded as $5a + 5b$, then worked.
$(5 \times a) + b$ can be simplified to $5a + b$, then worked.

If $a = 6$, $b = 2$ are substituted in both expressions they become:

$5(6 + 2) = 5 \times 8 = 40$ and $(5 \times 6) + 2 = 30 + 2 = 32$

In algebra, first work anything that is inside brackets.

To **expand** brackets means to remove them by performing operations.
The **distributive rule** is used to expand brackets. Every term in the bracket must be multiplied or divided by the term outside the bracket. For example:

$7(6 + 3 - 1)$ It is not necessary to put the multiplication sign before the bracket.

$= 7 \times 8 = 56$ The bracket is worked first.

We can also write this as:

$(7 \times 6) + (7 \times 3) - (7 \times 1) = 42 + 21 - 7 = 56$ The 7 is distributed over each term.

The **vinculum** or dividing line functions like a bracket because it groups what is at the top or at the bottom. For example:

$\frac{6 + 4}{2} = (6 + 4) \div 2 = 10 \div 2$ $\frac{40}{12 - 8} = 40 \div (12 - 8) = 40 \div 4$

When you are working without a calculator brackets can be used to make your calculations much easier. Look at these examples:

$18 \times 1\,111$ is the same as: $18 \times (1\,000 + 100 + 10 + 1)$

$= 18\,000 + 1\,800 + 180 + 18 = 19\,998$

56×999 can be written as: $56 \times (1\,000 - 1)$

$= 56\,000 - 56 = 55\,944$

Worked examples

1. Expand the brackets:

   a. $6 + 2(a + b) = 6 + 2a + 2b$

   2(a + b) means $2 \times a + 2 \times b$

   b. $a(a - 2) = a^2 - 2a$

   a(a − 2) means $a \times a - a \times 2$

Exercise

1. Expand the following:
   a) $15(6 + k)$
   b) $3(a + b + c)$
   c) $26r + 4(p + 3 - q)$
   d) $\dfrac{3m - n + p}{2}$
   e) $6(g - h) + 5(j + k)$

2. Remove the brackets and simplify:
   a) $\dfrac{16 + 12 - 8}{4}$
   b) $3(a + b) + 2(a - b)$
   c) $m(2 + m) + m(2 - m)$
   d) $12(7 - a) + 14(a - 2)$
   e) $(9x + y + 3z) + 2(x + y + z)$

3. Use brackets to split one of the numbers, then use the distributive rule to calculate:
   a) 199×48
   b) 58×21
   c) 101×561
   d) $1\,002 \times 934$
   e) 98×365
   f) Use a calculator to check your answers.

Investigation

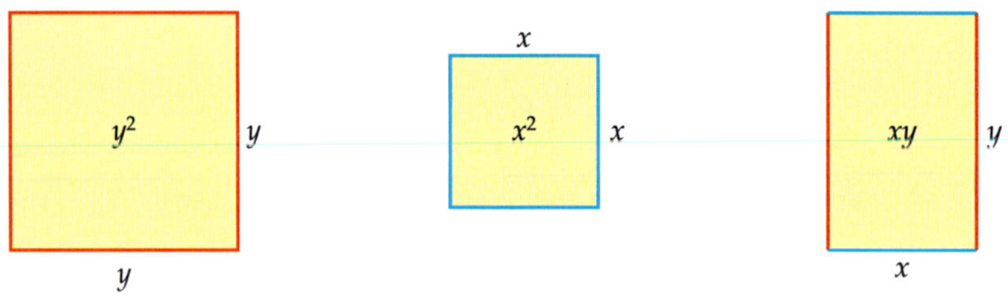

Use models of y^2, x^2 and xy to represent:

a) $3y^2 + x^2 + 2xy$

b) $4(3y^2 + x^2 + 2xy)$

Use your models to write another expression for part **b)**.

Expand the brackets in part **b)** and compare the results with that of your model.

Key words
Commutative
Order of operations

Unit 8.4 Order of operations in algebra

In this unit you will:

- **recognise that the order of operations for algebra is the same as for arithmetic**
- **use the correct order of operations to simplify expressions or calculate their values.**

There are set rules for the order in which the operations +, –, ×, ÷ and brackets must be worked. You learned about these in Chapter 2. We will revise them briefly here.

When there is addition and subtraction only, work in the order in which they come, from left to right. For example:

$6 - 2 + 1 = 4 + 1 = 5$

$8 + 3 - 4 = 11 - 4 = 7$

When there is multiplication and division only, work in the order in which they come, from left to right. For example:

$16 \div 4 \times 2 = 4 \times 2 = 8$

$12 \times 2 \div 8 = 24 \div 8 = 3$

When there are brackets, do the brackets first. For example:

$7 - (6 - 5) = 7 - 1 = 6$

$(5a + 13a) - 8a = 18a - 8a = 10a$

Multiplication and division are done before addition and subtraction. For example:

$6 + 8 \times 2 = 6 + 16 = 22$ and $16 - 3 \times 3 = 16 - 9 = 7$

$12 + 10 \div 2 = 12 + 5 = 17$ and $9 - 6 \div 3 = 9 - 2 = 7$

When terms exchange places and the results are the same the operation is **commutative**. For example:

$15 + 4 = 4 + 15$ $\quad 6 \times 3 = 3 \times 6$

Addition and multiplication are commutative.

$15 - 4 \neq 4 - 15$ $\quad 6 \div 3 \neq 3 \div 6$

≠ means 'not equal to'. Subtraction and division are not commutative.

To simplify expressions follow the **order of operations** you learned for numbers.

Worked examples

1. $9 \times 4 - (7 - 3)$

$= 9 \times 4 - 4$ (Bracket first)

$= 36 - 4$ (Multiplication next)

$= 32$

2. $7a + 6a - 3(a + 3a)$

$= 7a + 6a - 3(4a)$ (Bracket first)

$= 7a + 6a - 12a$

$= 13a - 12a$ (From left to right, the order in which + and − come)

$= a$

3. $18 - 12 \div (5 + 1) \times 3$

$= 18 - 12 \div 6 \times 3$ (Bracket first)

$= 18 - 2 \times 3$ (From left to right, the order in which × and ÷ come)

$= 18 - 6$ (Multiplication is done before subtraction)

$= 12$

Exercise

1 Simplify the following:

a) $65 - 42 + 31 - 15$

b) $4 + 3 \times 8 - 2$

c) $\frac{72 + 4}{2}$

Remember the vinculum works like a bracket.

d) $\frac{21 + 9}{7 + 3}$

e) $25 + 15 \div 3 - 5$

f) $17 + 32 \div 4(5 - 3)$

2 Simplify the following:

a) $15n + 2n - 12n - 4n$

b) $6m + 4m \times 2$

c) $27 - 8(p + 3)$

d) $85 + 5(m - 6)$

e) $3d + 12d \div 6d$

Investigation

Use the numbers 12, 4, 9, 3 (in the same sequence) with any of the four operations (+, −, ×, ÷) or brackets, to write five problems which have different answers.

For example: $12 - 4 + 9 \times 3 = 35$ $(12 - 4) \times 9 \div 3 = 24$

Looking back

1 Simplify:

a) $t + t - q + t - q - q$

b) $p + r - p + r + r - p$

c) $x + y + y - x - y + x + y$

d) $b - t - t - t$

e) $x + y - x - y$

f) $t + y - u - u + y - t$

2 Expand:

a) $6y$

b) $3z + 2y$

c) $7a + 3b + 2c$

d) $2s$

e) $4 \times 3y^3$

f) $x + x^3 + 4 \times 2$

3 Simplify the following expressions:

a) $6 + 2y - 3 + 4y$

b) $16 + t - 2s + 12$

c) $3w - 6v + 4v - 100$

d) $79t + 50 - 30t - 3$

4 Substitute the given values for a to find the value of p in the expression $p = 20a - 6$.

a) $a = 3$ **b)** $a = 4$

c) $a = 10$ **d)** $a = 5$

5 Expand the brackets and simplify:

a) $2(5 + 2c - 4)$ **b)** $8(r - 5) + 12$

c) $6(4r - 2t + 10r)$ **d)** $6a - \frac{12}{3} + 12a - \frac{15}{3}$

6 How many terms are in the following algebraic expressions?

a) $x + 2y$ **b)** abc **c)** $4xy + 2$

d) $3c + d + b$ **e)** $ab + cd$ **f)** $8a + 2b + c$

g) $x - 8$ **h)** $2c$ **i)** $b + b + b + b$

j) $9 \times u$ **k)** $3 \times 3 \times n$ **l)** $a + c + 3 + 7$

m) $(x + 3y)$ **n)** $x + 3y$ **o)** $3(a + b)$

p) $(9x + c) - (4x + d)$ **q)** $3x + 4y + (m + n)$

7 Simplify:

a) $2a \times 3 + 4a \times 2$ **b)** $5m \times 2 + m \times 3$ **c)** $6b \times 4 + 6b \times 5$

d) $2a + 3a \times 6$ **e)** $5q + 4 \times 3q$ **f)** $10k + k \times 5$

8 If $a = 2$, $b = 1$ and $c = 3$, find the value of:

a) $a + b + c$ **b)** $2a + 2b + 2c$ **c)** $2(a + b + c)$

d) $a \times a$ **e)** abc **f)** $2abc$

g) c^2 **h)** $a(b + c)$ **i)** $b(a + c^0) + c(a + b)$

j) $\frac{2c}{a}$ **k)** $\frac{2c}{3b}$ **l)** $3(2a + 2b + 4c)$

m) $5a - 2b$ **n)** $3c - 2a - b$ **o)** $a + b^0(a + c)$

Remember: $x^0 = 1$.

9 Write the following as algebraic expressions. When you have done this, calculate their value if $x = 6$.

a) A number 6 more than x.

b) Three times x increased by 4.

c) The square of x decreased by 7.

d) Twice the product of x and 12 increased by 2.

e) A number which is 12 less than three times x.

Project

This shape has

- 8 dots on its boundary
- 2 dots inside the boundary
- an area of 5 square units.

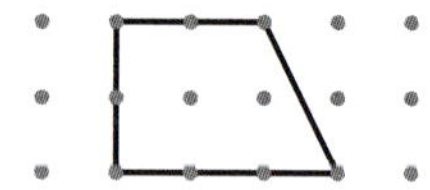

1 **a)** Copy and complete the table for each of these shapes.

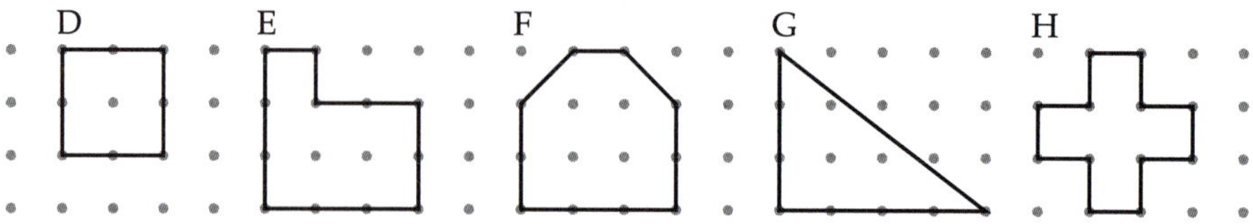

Shape	Number of dots on boundary (b)	Number of dots inside boundary (i)	$\frac{1}{2}b + i$	Area (A) in square units
D	8	1		
E				
F				
G				
H				

b) Draw some shapes of your own and add the information to your table.

c) Look at the last two columns in your table. What do you notice?

d) Copy and complete the formula: $A = \frac{1}{2}b + i - \square$.

This is called Pick's formula.

2 Use the formula to find the area of each of these shapes.

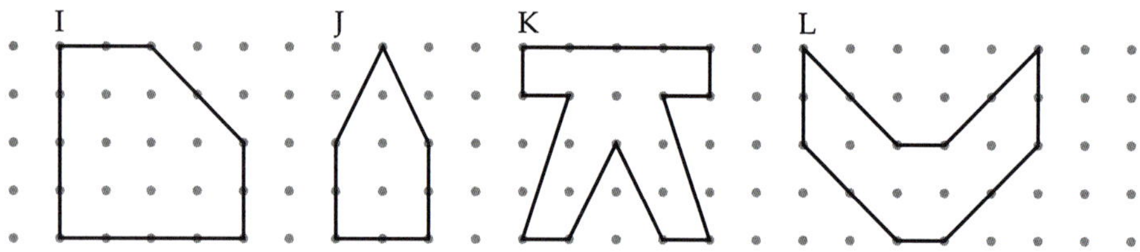

3 Use Pick's formula to find the shaded area in each of these shapes.

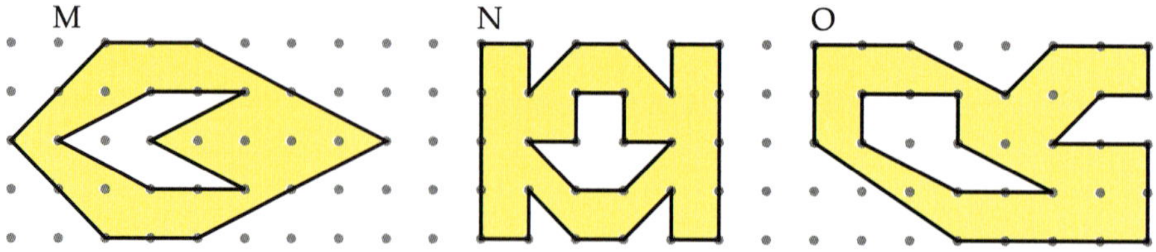

Project

1 In a new holiday resort in Tobago, the path to the front door of each villa has a row of grey tiles between two rows of white tiles.

This path has 3 grey tiles and 6 white tiles.

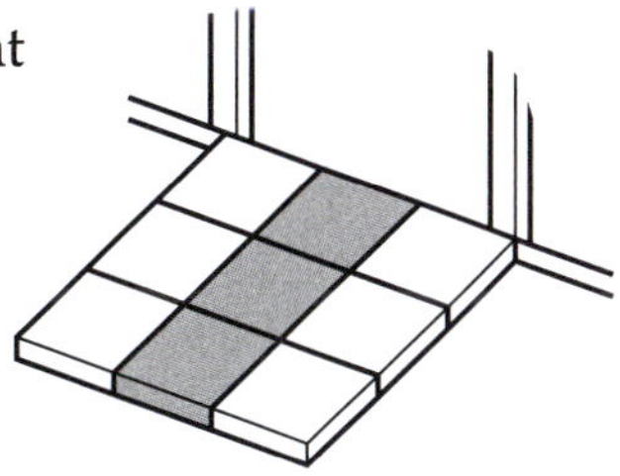

a) Copy and complete this table for paths with
- 4 grey tiles
- 5 grey tiles
- 10 grey tiles
- 20 grey tiles.

Number of grey tiles	Number of white tiles
3 →	6

b) Describe how to find the number of white tiles when you know the number of grey tiles.

c) How could this information help builders?

2 Each pavement is made from slabs and border tiles. Border tiles are placed along one side and at each end of the pavement. This pavement has 5 slabs and 7 border tiles.

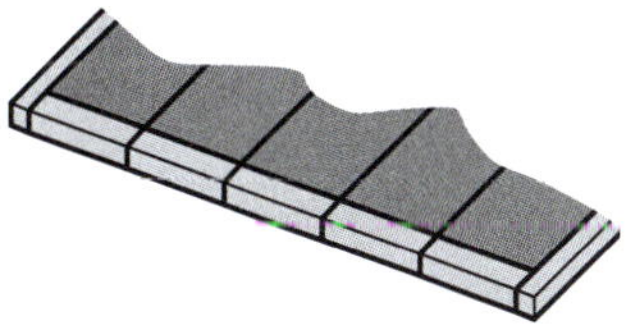

a) Copy and complete this table for
- 3 slabs
- 4 slabs
- 10 slabs
- 20 slabs.

Number of slabs	Number of border tiles
5 →	7

b) Explain how to find the number of border tiles when you know the number of slabs.

3 Two houses side by side have 1 joining wall.

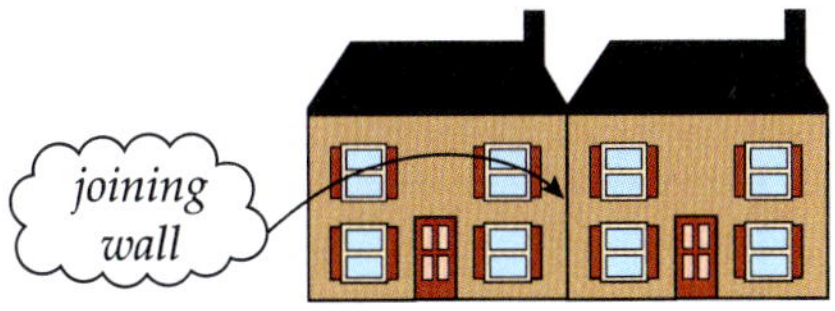

Three houses side by side have 2 joining walls.

a) Copy and complete this table for
- 4 houses
- 5 houses
- 10 houses
- 30 houses.

Number of houses	Number of joining walls
2 →	1
3 →	2

b) Describe how to find the number of joining walls when you know the number of houses.

Revise and consolidate

1. Rewrite each length by converting it to the units given:

   a) 45.9 km = ____ m

   b) 58 000 cm = ____ km

   c) 9.2 m = ____ cm

   d) 42 mm = ____ cm

   e) 980 m = ____ km

   f) 0.006 73 km = ____ mm

   g) 564.7 cm = ____ m

   h) 8.52 cm = ____ m

   i) 609 mm = ____ cm

   j) 6 700 m = ____ km.

2. Find the perimeter of the following shapes. Give all answers in cm.

   a)
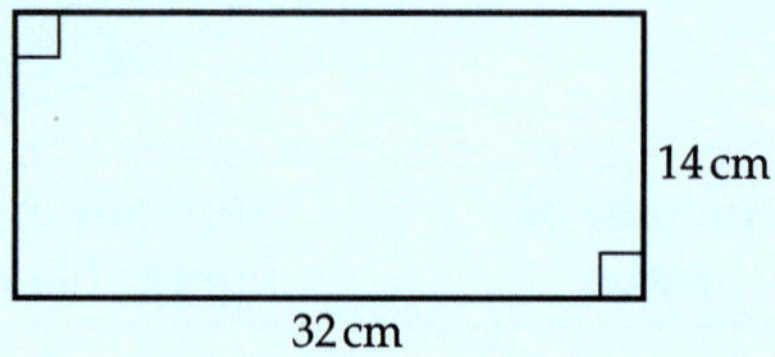

   b)
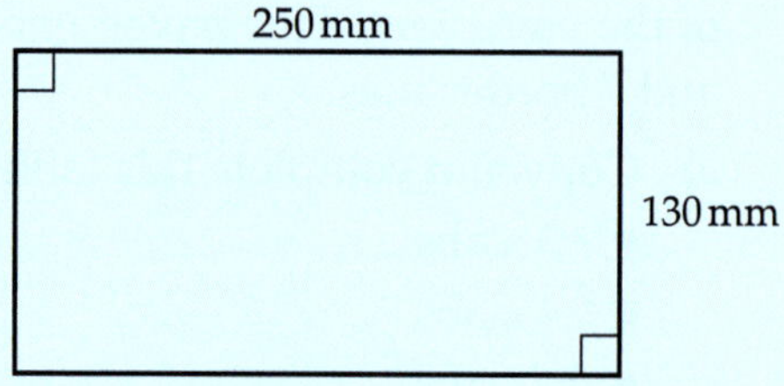

   c)
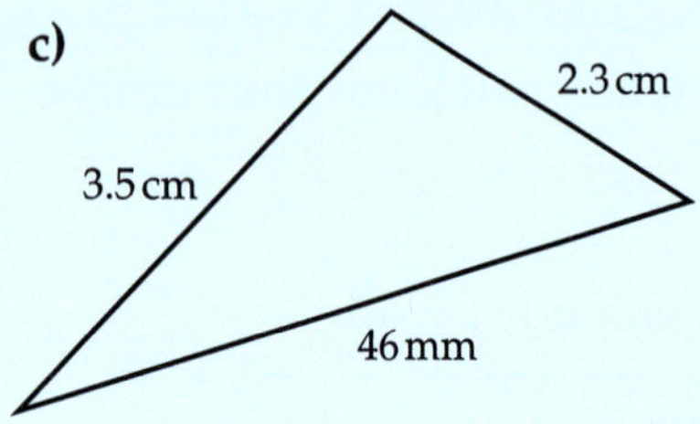

   d)
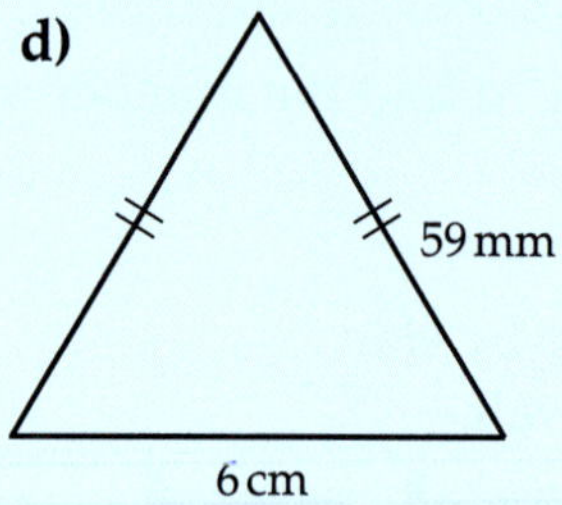

   e)
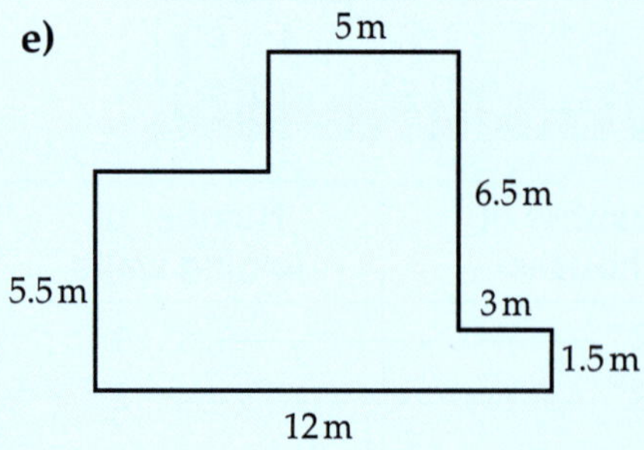

   f)
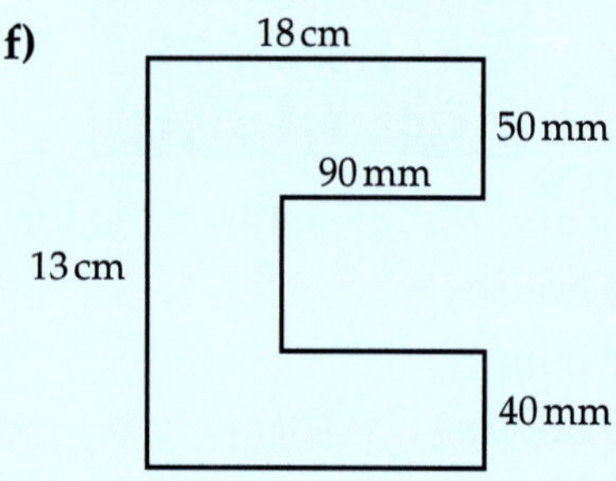

3. What is the perimeter of a squash court 9.75 m long and 6.4 m wide?

4 Complete the following statements by inserting an approximate measurement and appropriate units.

a) The length of a paper clip is about ______.

b) The height of our classroom ceiling is about ______.

c) A cricket ball weighs about ______.

d) A bucket holds about ______.

e) It takes me about ______ to do my homework.

f) Today the temperature must be about ______.

5 Say whether the following statements are true or false:

a) $6y$ is a term.

b) $7y - 9$ is an equation.

c) $y = 7x$ is a term.

d) ab is a term.

e) $r = t + 9$ is an expression.

f) $cd + 4$ is an expression.

g) $w = 2 - 5b$ is an equation.

h) rs and sr are like terms.

6 Write expressions to represent the following statements:

a) I take any number and add five to it.

b) I take any number and add 14 to it.

c) I multiply any number by six and then add 7 to it.

d) I take any number, multiply it by eight and then add 12.

7 If $x = 2$, find the value of $4x - 1$.

8 If $x = 7$, find the value of $x - \frac{3}{7}$.

9 Mark's collection of three types of stamps is $45t + 6s + 22p$, John's collection is $55s + 19p$. What is their total collection?

10 Find the total length of wire needed to make the following shapes.

a)

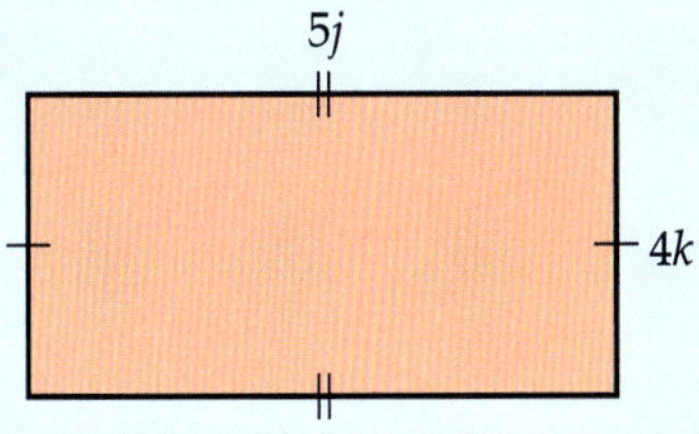

b)

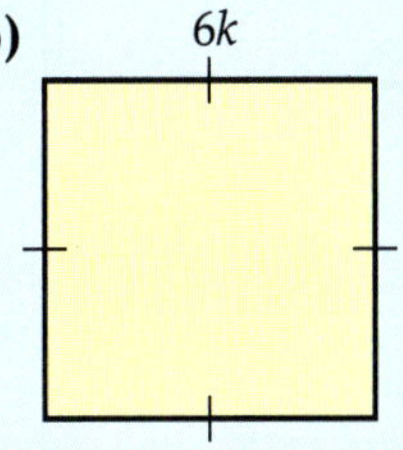

c)

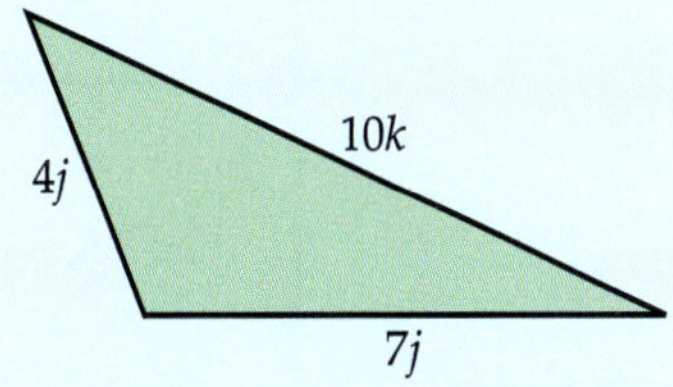

11 The cost of items in a sale is $\$(10 + \frac{1}{2}p)$; p is the price before sale. What is the cost of the following items on sale?

a)
b)
c)
d)

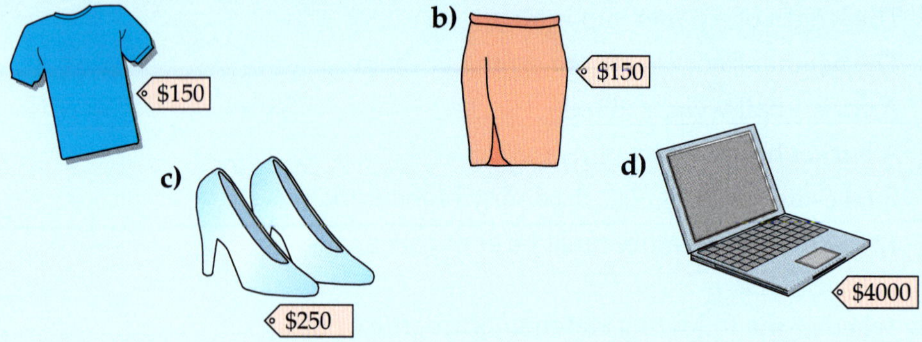

12 Find the area of a rectangle of length 54 cm and width 34 cm.

13 Find the area in km² of a park $3\frac{1}{2}$ km long and 900 m wide.

14 A yacht has a triangular sail which is 2.1 m high and 1.6 m long. Sketch the sail and calculate its area.

15 Find the area of Jason's diamond shaped kite:

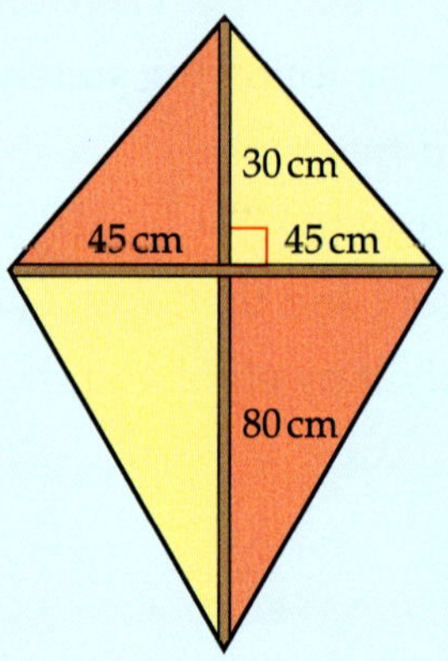

16 Find the area of this sugar cane farm using the plan provided.

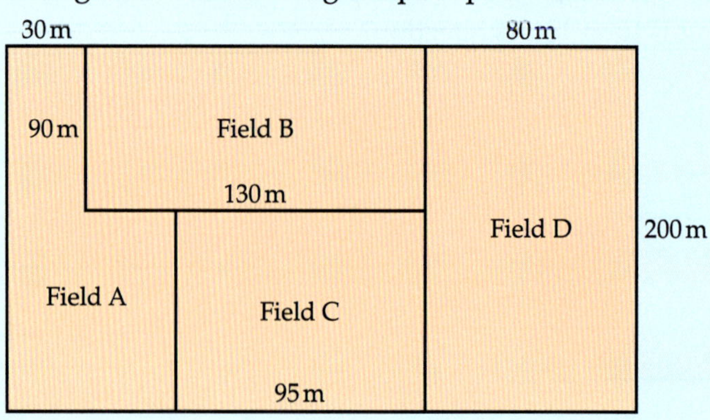

Key words

Equal
Expression
Equation
Formula/formulae

Chapter 9

Equations and formulae

This chapter will help you understand how to use formulae and equations from algebra to solve real-life problems. You will apply your knowledge of algebra to represent operations using flow charts and create your own formulae and equations.

Unit 9.1 Generating equations and formulae

In this unit you will:

- associate formulae with real-life situations
- develop simple equations to describe and solve word problems.

You can write a mathematical sentence to describe the perimeter of these shapes:

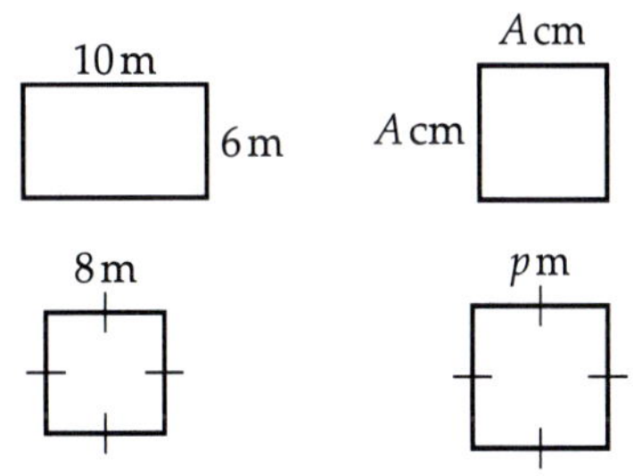

Mathematical sentences must contain an **equal** sign (=).
'=' is read 'is equal to', although we can read $4 + 3 = 7$ as '4 plus 3 equals 7'

An **expression** is not a mathematical sentence because it does not contain an equal sign.
For example, $p + 6$ is an expression.

You have already learnt about **equations**.
An equal sign means that both sides of the equation have the same value.
For example, $x + 4 = 8$.

A **formula** is a general equation that links variables. For example, we can write a formula for finding the area of any rectangle. The formula is:

$A = L \times B$ where L is the length and B is the breadth. A stands for area.

The plural of formula is **formulae**. Formulae are used in everyday life. For example, in Trinidad, Value Added Tax (VAT) on items is 15% of the cost. The formula for calculating this is:

$T = 0.15C$ where T is the tax and C is the cost of the item.

Worked examples

1. Think of a number; add 24 and the result is 30. Write an equation using r for the number.
2. Use any variable to write the following equation:
   Mr Joe is 6 years older than Mrs May. The sum of their ages is 56.

1. $r + 24 = 30$
2. Let Mr Joe's age be m years, Mrs May is $m - 6$ years.
   The sum of their ages is $m + m - 6 = 56$.

3. a. The cost of an adult ticket for a concert is $\$m$. The cost for children under 12 years is half price. Write an equation for the total cost (C) for an adult and a child.

b. Find the cost for a child and adult on a Saturday when $m = \$450$

c. Find the cost for a child and adult on a Sunday when $m = \$400$

3. a. $C = m + \frac{m}{2}$

b. On Saturday $C = \$450 + \frac{\$450}{2} = \$675$

c. On Sunday $C = \$400 + \frac{\$400}{2} = \$600$

Exercise

1 Put the following under the best heading:
Simple equations, Expressions, Formulae.

a) $P = 2L + 2W$ — Perimeter of rectangles.

b) $7rs + 5 - 4s$

c) $F = \frac{9}{5}C + 32$ — Changing from Centigrade to Fahrenheit.

d) $5r + 24 = 7r - 1$

e) $45 + 4p = 17$

f) $mx + c$

2 **a)** A worker gets $\$c$ per weekday and $\$c + r$ on weekends. Write an equation for a week's pay if a man works six days, including one weekend day.

b) How much was his pay if $c = \$300$ and $r = \$200$?

c) How much was he paid if $c = \$550$ and $r = \$100$?

3 Write a simple equation for each of the following. Use y for the variable.

a) 6 times a number is 24

b) 5 less than a number is 72

c) 13 more than a number is 29

d) 8 added to twice a number gives 22

e) Take 35 from a number; the result is 95

4 Pat is w years old. Pete is 4 years younger.

a) Write an expression for Pete's age.

b) Write an expression for the sum of their ages.

c) Write an equation if the sum of their ages is 30.

Investigation

The formula for calculating the cost (C) of a table is $C = L + M$. L is the cost for labour and M the cost for material. Which of the following tables costs the least?

a) Labour \$5 500, material \$4 760.

b) Labour \$5 000, material \$5 170.

c) Labour \$4 000, material \$5 840.

Key words
Flow chart
Input
Output
Variable
Inverse operation

Unit 9.2 Using flow charts to solve equations

In this unit you will:

- use flow charts to develop and solve simple equations.

A **flow chart** (or flow diagram) shows the instructions to be followed in order to get results.

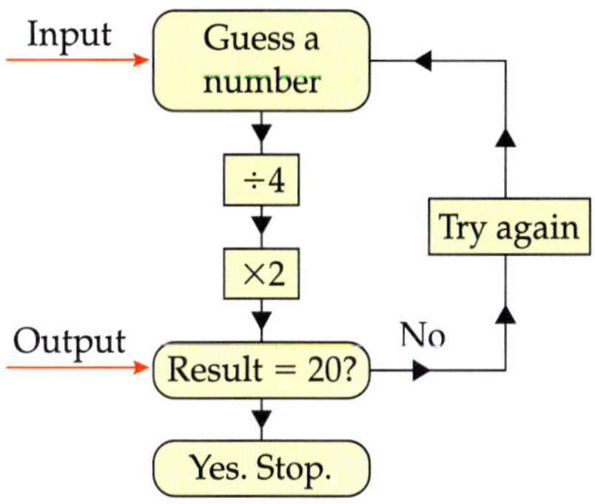

A flow chart starts with an **input**.
The end result is called an **output**.

In mathematics we often need to find an unknown number or **variable**: $1 + \square = 5$

In the example above, the number we are looking for is 4. $1 + \boxed{4} = 5$

We can write this as $\square = 4$.

We can use a mapping diagram to show how one number moves to another number by using a rule. The diagram on the right shows that the rule to get from the numbers above to the numbers below is 'add 4'.

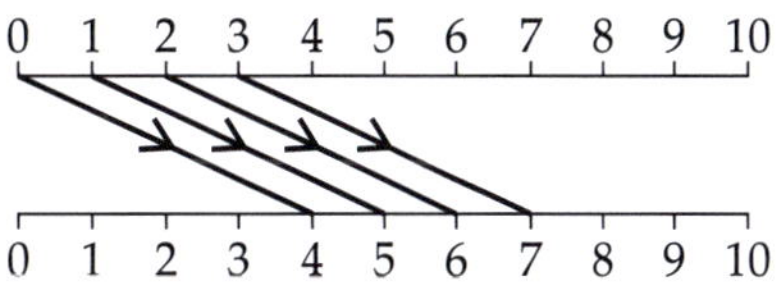

We can also represent this information using a simple flow chart.

Input $\rightarrow \boxed{+4} \rightarrow$ output.

Worked examples

1. Find the number represented by $\square$:

a. $\square \div 2 = 12$ **b.** $\square \times 5 = 65$

2. Look at this flow chart: input $\rightarrow \boxed{+5} \rightarrow$ output

a. If the input is 8, what is the output? **b.** If the output is 6, what is the input?

1. a. $\boxed{24} \div 2 = 12$
$\square = 24$
Check: $24 \div 2 = 12$

Always check your answer.

b. $\boxed{13} \times 5 = 65$
$\square = 13$
Check: $13 \times 5 = 65$

Count up in fives till you know how many fives make 65.

2. a. $8 \rightarrow \boxed{+5} \rightarrow 13$
The output is 13.

b. $? \rightarrow \boxed{+5} \rightarrow 6$
The input must be 1. Check: $1 + 5 = 6$ ✓

One way of finding the missing numbers is to work backwards. To work backwards, we use **inverse operations**. Remember that addition is the inverse of subtraction and multiplication is the inverse of division.

Worked examples

1. Show the following information on a flow diagram. Use a variable to represent the unknown: 'Think of a number, divide it by 4 and then add 7. The result is 10.'
2. Write this as an equation.
3. Work backwards to find the input and check your result using the equation.

1. $x \rightarrow \boxed{\div 4} \rightarrow \boxed{+7} \rightarrow 10$ **2.** $\frac{x}{4} + 7 = 10$

3. Input is 12. $(10 - 7) \times 4 = 12$

Check in **2.** $\frac{12}{4} + 7 = 3 + 7 = 10$

Exercise

1 What is the rule that links the following inputs to the outputs? Only one operation (+, −, × or ÷) is used in each.

Input	Output
3	15
4	20
5	25

2 Fill in the correct operation sign (+, −, × or ÷):

a) 8 ☐ 4 = 2 **b)** 6 ☐ 12 = 72 **c)** 14 ☐ 7 = 2

d) 9 ☐ 15 = 24 **e)** 18 ☐ 2 = 9 **f)** 48 ☐ 8 = 40

3 Copy the diagrams and insert numbers to link the input with the output.

a)

Input	5
	−?
Output	1

b)

6
+?
7

c)

?
÷18
3

d)

17
+?
30

e)

?
×4
52

4 Write an equation for the following inputs and outputs.

a)

Input	12	13	14	m
Output	6	7	8	n

b)

Input	1	2	5	x
Output	11	12	15	y

5 Write the equation for each of the following flow diagrams.

a) $\rightarrow \boxed{x} \rightarrow \boxed{\div 2} \rightarrow \boxed{21}$

b) $\rightarrow \boxed{y} \rightarrow \boxed{+g} \rightarrow \boxed{z}$

c) $\rightarrow \boxed{2x} \rightarrow \boxed{y}$

d) $\rightarrow \boxed{2x} \rightarrow \boxed{-10} \rightarrow \boxed{22}$

Key words
Equation
Balance
Inverse operation
Solve

Unit 9.3 Solving simple equations

In this unit you will:

- solve equations.

You should remember that an **equation** is when two sides of a mathematical sentence are equal. An equal sign means that both sides have the same value.

Look at these sets of scales. We can **balance** the scales by adding or subtracting mass pieces:

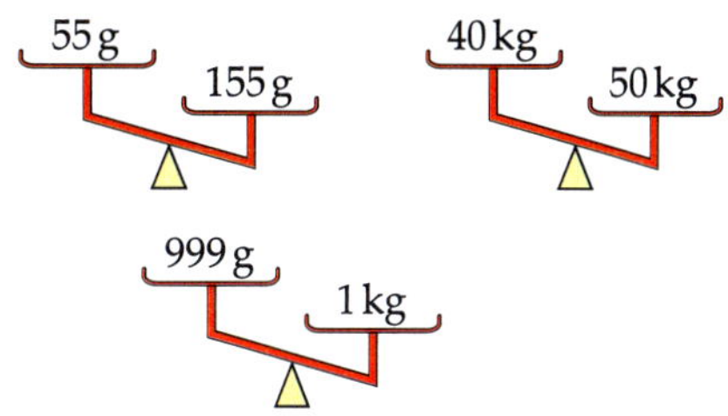

You can think of the equal sign as a level scale – both sides of the equation must remain balanced in order for the scale to stay level. This means that if you do something to one side of the equation, you have to do the same to the other side.

We can also use **inverse operations** to **solve** equations. Look at these different ways of writing pairs of equations:

$6 + 4 = 10$	$8 - 3 = 5$	$12 \div 3 = 4$	$5 \times 2 = 10$
$6 = 10 - 4$	$8 = 5 + 3$	$12 = 4 \times 3$	$5 = 10 \div 2$

Worked examples

Solve the equations:

a. $x - 4 = 10$ **b.** $48 = 8p$ **c.** $\frac{k}{4} = 7$

a. Using **inverse operations**: or by **balancing** the equation:

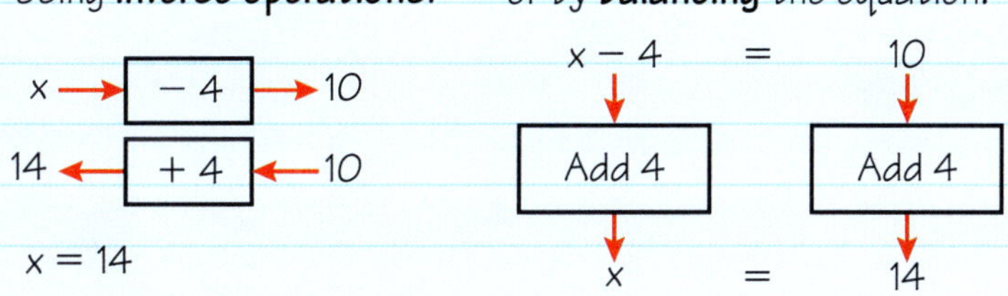

b. $48 \div 8 = p$ (Balance the equation by dividing both sides by 8.) $6 = p$

c. $k = 7 \times 4$ (Multiply both sides by 4 to balance the equation; $\frac{k}{4} \times 4 = k$.) $k = 28$

Remember to check your solutions by inserting the number in place of the unknown to make sure the equation balances.

Exercise

1 Solve the following equations. Show your inverse operations.

a) $q + 6 = 29$ **b)** $m - 2 = 17$ **c)** $4p = 32$ **d)** $48 = 6x$

e) $8 = x - 1$ **f)** $84 = 12x$ **g)** $r \div 5 = 45$ **h)** $50h - 100 = 0$

i) $\frac{y}{15} = 1$ **j)** $65 = 15 + d$

Sometimes equations may require more than one step to solve them.

Simplify when more than one term has the unknown.

Balance the equation to group the terms with the unknown on one side of the equal sign and place the numbers only on the other side.

Always to do the same thing to both sides of the equation so that it stays balanced.

Worked examples

1. $5g + 16 - 6g = 26 - 2g$

$5g + 16 - 16 - 6g = 26 - 16 - 2g$ (Subtract 16 from both sides.)

so $5g - 6g = 10 - 2g$

so $5g - 6g + 2g = 10 - 2g + 2g$ (Add $2g$ to both sides.)

so $7g - 6g = 10$

so $g = 10$

Check by putting 10 in place of g on both sides.

Left hand side (LHS): $5(10) + 16 - 6(10) = 50 + 16 - 60 = 66 - 60 = 6$

Right hand side (RHS): $26 - 2(10) = 26 - 20 = 6$

2. $a + 7a - 8 + 3a = 4a + 20$

$a + 7a - 8 + 8 + 3a = 4a + 20 + 8$ (Add 8 to both sides.)

so $a + 7a + 3a - 4a = 4a - 4a + 28$ (Subtract $4a$ from both sides.)

so $a + 7a + 3a - 4a = 28$

so $11a - 4a = 28$

so $7a = 28$ (Simplify both sides.)

so $\frac{7a}{7} = \frac{28}{7}$ (Divide both sides by 7)

so $a = 4$

Check: LHS: $4 + 7(4) - 8 + 3(4) = 4 + 28 - 8 + 12 = 36$

RHS: $4(4) + 20 = 16 + 20 = 36$

3. $\frac{9t - 10 + 2t}{3} = (t + 2)$

$\not{3}\left(\frac{9t - 10 + 2t}{\not{3}}\right) = 3(t + 2)$ (Get rid of the fraction by multiplying both sides by 3.)

so $9t - 10 + 2t = 3t + 6$

so $9t - 10 + 10 + 2t = 3t + 6 + 10$ (Add 10 to both sides.)

so $9t + 2t = 3t + 16$

so $9t + 2t - 3t = 3t - 3t + 16$ (Subtract $3t$ from both sides.)

so $11t - 3t = 16$

so $\frac{8t}{8} = \frac{16}{8}$ (Divide both sides by 8.)

so $t = 2$

Check: LHS: $\frac{9(2) - 10 + 2(2)}{3} = \frac{18 - 10 + 4}{3} = \frac{12}{3} = 4$

RHS: $2 + 2 = 4$

You may find shorter methods to solve equations by skipping some steps. However, it is important to understand what is happening, so we have shown <u>all</u> the steps.

Exercise

1. Solve the following equations. Check your answers.
   a) $2x - 5 = 9$ b) $5y + 4 = 39$ c) $y + 30 = 7y$ d) $5 + 4m = 18$ e) $11 = \frac{m}{2}$

2. Simplify and solve the following.
   a) $6y - 2y - 3y = 16$
   b) $10f - 18 = 12f - 3f$
   c) $3a + 7 + 4a - 37 = 2a$
   d) $6x - 3x + 12x - 45 = 4x - 1$
   e) $10 - 4 + 8n = 13n + 4n - 3$

3. Solve the following equations and check your answers are correct by substituting.
   a) $3x + 2 = 20$
   b) $5x - 7 = 28$
   c) $28 = 3x + 4$
   d) $5x + 9 = 4$
   e) $4x + 15 = 7$
   f) $2x + 1 = 5$
   g) $2x + 2 = 6$
   h) $2b + 8 = 12$
   i) $2x - 2 = 8$
   j) $3 + 2f = 13$
   k) $2x + 1 = 2$
   l) $4g + 6 = 22$
   m) $2 + 3h = 5$
   n) $5x - 1 = 2$

4. Solve each of the following equations and check your answers.
   a) $\frac{x}{2} + 1 = 3$
   b) $\frac{a}{3} - 4 = 2$
   c) $\frac{x}{7} + 7 = 7$
   d) $\frac{d}{4} - 2 = 6$
   e) $1 + \frac{d}{10} = 5$
   f) $3 + \frac{y}{8} = 9$

5. Simplify and solve.
   a) $\frac{4x - 1}{3} = 5$
   b) $\frac{x + 5}{3} = 11 - x$
   c) $2(2 - x) = 3(x + 3)$
   d) $5(p - 1) - (1 - 2p) = 8$
   e) $7 - r = \frac{2r - 7}{5}$
   f) $2y - 3(y - 1) = 15 - 4y$

6. Three times a certain number decreased by 2 is 7. What is the number?

7. When half a certain number is subtracted from 11, the result is 5. What is the number?

8. The sum of three consecutive whole numbers is 27. What are the numbers?

9. Ravi is two years older than Javid and four years older than Kwesi. Together their ages add up to 45. How old is Javid?

10. The length of a rectangle is three times its breadth. If the perimeter of the rectangle is 88 cm, how long is each side?

Looking back

Use your knowledge of algebra and arithmetic to solve the clues. Then copy and complete the crossword puzzle below:

Clues

Across

1. $x + 120 = 200$
2. $y + 95 = 400$
4. The number of days in a year.
6. I choose a number, subtract 60 and get 1995. What number did I choose?
7. I multiplied this number by 11 and got 121.
8. $3x + 1 = 37$
9. $\frac{m}{3} = 111$
10. $1.5x = 46.5$
12. I choose a number, add 7 and get 21. What was the number?
13. Double 70.
14. $15\,000 - x = 8\,000$
15. $2x + 5 = 107$

Down

1. I choose a number, divide it by 2 and get 401. What number did I choose?
2. $\frac{r}{7} = 5$
3. I add 40 to this number and get 551.
4. $2x + 10 = 80$
5. $10a = 650$
7. To get this number, you add 10 000 and 2 000 and then add 300, then add 50, and subtract 5.
9. $3x - 1\,000 = 20$
11. I am 5 more than 96.
12. If I divide 34 by this number, I get 2.

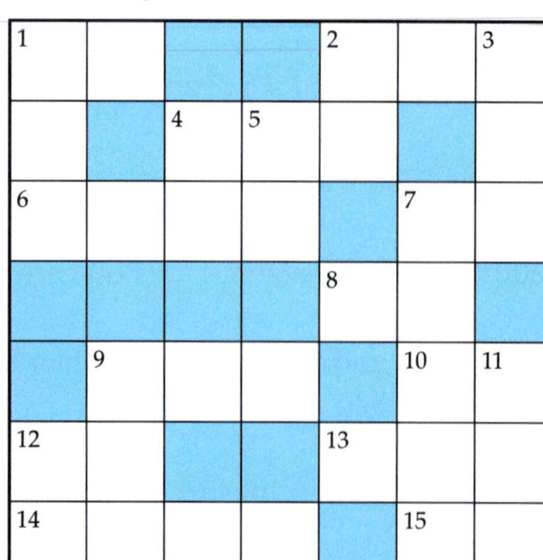

Project

Isha designs road systems. All the streets in this new town are to be straight lines. At each crossroads a control box will be needed to operate the traffic lights.

Three roads **could** cross in the following ways.

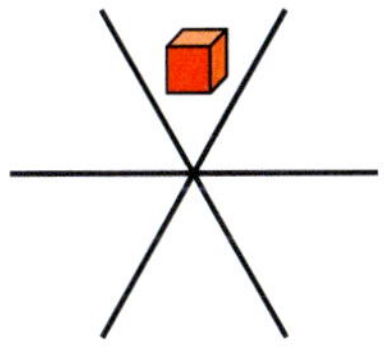

1 control box is needed.

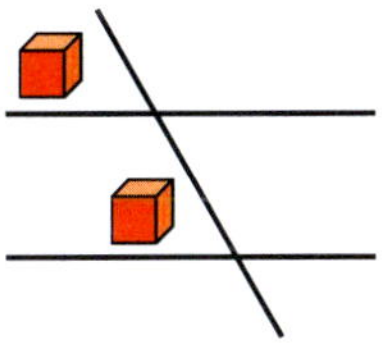

2 control boxes are needed.

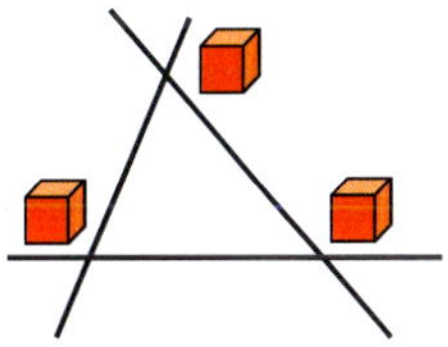

3 control boxes are needed.

The **maximum** number of control boxes needed for 3 roads is **3**.

1 Draw a diagram to show the maximum number of control boxes needed for a design with

a) 2 roads **b)** 4 roads.

2 Copy this table. Record the maximum number of control boxes needed for designs with 2, 3 and 4 roads.

Number of roads	1	2	3	4	5	6
Maximum number of control boxes	0					

3 **a)** What do you think is the maximum number of control boxes needed for 5 roads? Check by drawing a diagram.

b) Complete your table.

4 Isha's design has 8 roads in the centre of the new town. What is the greatest number of control boxes needed for the town centre?

5 **a)** Look at the diagram for 3 roads. There are 2 control boxes on each road. How many control boxes are there altogether?

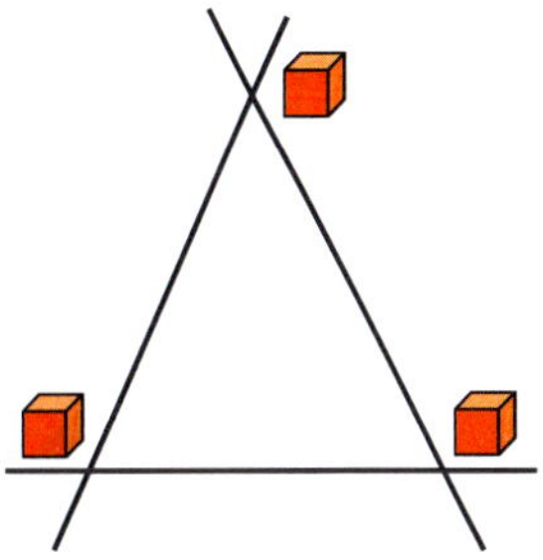

b) Look at **your** diagram for 4 roads. How many control boxes are there on **each** road? How many control boxes are there altogether?

c) Repeat part **b)** for 5 roads.

d) Describe how Isha could find the maximum number of control boxes when she knows the number of roads.

6 There are to be 20 roads in the new town. What is the greatest number of control boxes needed?

7 Write a formula to find the maximum number of control boxes needed for a design with n roads. Check your formula.

Chapter 10

Consumer arithmetic

This chapter deals with the value and functions of money in our daily lives. Informed consumers understand how money works, appreciate its value and can perform the four basic operations on money amounts. This chapter also equips you for real life by teaching skills and calculation methods that you can use to solve problems related to buying and selling, value for money, discounts, taxes, hire purchase, profit and loss, foreign exchange and simple interest.

Key words
Exchange
Value
Saving
Wealth
Currency
Denominations

Unit 10.1 Understanding money

In this unit you will:

- understand the value and functions of money
- identify currencies used by different countries
- investigate banking and saving schemes.

Today we all use money in the form of coins and notes. We may **exchange** our money for goods or services and we receive money in exchange for our labour or services. The Chinese were the first to use paper money in the 10th century. Coins were heavy and so people began to store their coins with merchants in exchange for a paper receipt. The receipts soon became used as currency because although the paper itself had no **value**, it could be exchanged for metal coins and this gave it value. This exchange was probably also a form of **saving** and the term 'merchant banker' probably arose from this practice.

Money is also seen as a store of **wealth** – people who have large amounts of money are considered wealthy.

The money of a country is called its **currency**. Most currencies have notes and coins in different **denominations**. Different currencies also have different values. The value of a currency is linked to many different economic factors and it can change over time.

The information below gives you equivalent values (for 2005) of different Caribbean currencies against the US dollar.

TT$6.30

BD$1.99

EC$2.70

Ja$61.63

BSD$1

US$1.00

Sometimes countries decide to use a common currency. For example, the East Caribbean dollar is used by Anguilla, Antigua and Barbuda, Dominica, Grenada, Montserrat, St Kitts and Nevis, St Lucia, and St Vincent and the Grenadines.

Exercise

1. Write down the denominations of notes and coins used in Trinidad and Tobago.

2. Look at the diagrams carefully and answer the questions that follow:

The price of a weekly shopping bag in four different currencies.

Selina
Barbados
BD$2 985

Rafael
Jamaica
Ja$92 110

Agatha
St Lucia
EC$4 005

Ishmael
Trinidad
TT$9 422

These people are all bank tellers. They earn a monthly salary in exchange for their labour.

a) Who earns more dollars, a bank teller in Jamaica or a bank teller in St Lucia?

b) Which of the bank tellers earns the least dollars? Does this mean he or she is poorer than the others? Explain why or why not.

c) The price of a bag of groceries is higher in Jamaica than in the other countries. Is it really more expensive in Jamaica?

d) What fraction of each person's monthly salary is spent on a week's shopping?

e) What does this tell you about the cost of living in each country?

Investigation

What is the currency of:

- Bahamas
- British Virgin Islands
- Cuba
- Guadeloupe
- Haiti
- Netherlands Antilles?

Find out about different savings schemes offered by local banks. Try to find out:

- what different savings accounts are offered
- the cost of opening and running different accounts
- the services offered by different accounts
- how you open an account.

Discuss why people use banks to store their money.
Also discuss why some people prefer not to use banks.

Today, banking is computerised and people can draw money from their accounts using machines like the one opposite.

- Find out how these machines work.
- What are the advantages of ATM machines?
- Can you think of any disadvantages of ATM machines?

Unit 10.2 Problems involving percentages

Key words
Percentage

In this unit you will:

- calculate percentages of numbers, quantities and measures.

We can find a **percentage** of an amount in different ways:

a) Without using a calculator, find some useful percentages, for example, 50%, 10%, 1% and use combinations of these.

For example: to find 21%, double 10% and add 1%; to find 63%, find 50% then 10% and then 3 times 1%.

b) Converting the percentage to a common fraction and multiplying by the amount.

For example, find 25% of 400.

$25\% = \frac{25}{100}$

$\frac{25}{100} \times \frac{400}{1} = 100$

c) Using a calculator and the % key.

For example, to find 21% of 450, press 4 5 0 × 2 1 % =

Worked examples

1. A bag contains 75 mangoes. 20% of the mangoes are green. How many mangoes are ripe?
2. $4\frac{1}{4}$% of people who had to attend a conference were absent. If 17 people were absent, how many people were supposed to attend the conference?

1. $20\% \text{ of } 75 = 20\% \times 75 = \frac{20}{100} \times 75 = 15$ **or** $100\% - 20\% = 80\%$

   $75 - 15 = 60$ $\quad 80\% \times 75 = \frac{80}{100} \times 75 = 60.$

   So 60 mangoes are ripe.
2. $4\frac{1}{4}$% of people represents 17

   1% represents $17 \div 4\frac{1}{4} = 17 \times \frac{4}{17} = 4$

   100% represents $4 \times 100 = 400$ people. So 400 people were meant to attend.

Exercise

1 Calculate the following amounts:

a) 10% of \$140 **b)** 20% of \$75 **c)** 50% of 170 kg **d)** 60% of 85 m
e) 25% of 240 kg **f)** 75% of 6 m **g)** 90% of \$1 200 **h)** 40% of \$600
i) 30% of 40 kg **j)** 15% of 80 kg

2 Calculate the following amounts:

a) 11% of 70 km **b)** 21% of 160 m **c)** 15% of 32 g **d)** 51% of \$170
e) 24% of 120 **f)** 30% of 15 **g)** $6\frac{1}{2}$% of 1 000 **h)** 15% of 48
i) $12\frac{1}{6}$% of 900

Use a calculator to check your answers.

3 There are 128 grapefruits in a bag. $12\frac{1}{2}\%$ of them are rotten.

a) How many rotten grapefruits are there?

b) What percentage of grapefruits is not rotten?

4 Copy and complete this table to show different percentages of \$600, \$1 400 and \$3 200.

	\$600	**\$1 400**	**\$3 200**
20%			
35%			
12%			
51%			
42%			

5 40% of the students in Grade 3 are boys. There are 39 girls.

a) What percentage of Grade 3 students are girls?

b) How many students are in Grade 3?

c) How many boys are there?

6 A cyclist has travelled a distance of 32 km, which is equivalent to $33\frac{1}{3}\%$ of his journey. What is the total distance that he has to travel?

7 The distance from Jenny's home to her school is 50 km. On Tuesday, Jenny took the bus for 62% of the way, then walked a distance of 7 km and then got a ride to school with one of her classmates for the remainder of the way.

a) What percentage of the total distance did Jenny walk?

b) How far did Jenny ride with her classmate?

8 Victoria Johnson obtained the marks opposite in her mid-term examinations in four subjects:

Mathematics	$\frac{42}{70}$
Spanish	$\frac{71}{100}$
English	$\frac{36}{45}$
Science	$\frac{32}{60}$

a) Convert each mark to a percentage.

b) Which subject did she perform the best in?

c) Brenda Allen obtained $\frac{56}{84}$ for her mid-term examination in mathematics. Which of the two students had the better mark in their mathematics mid-term examination?

9 Nneko has \$35.72. She earns 2% of this amount. How much did she earn in cents?

Investigation

☐% of ☐ = ☐

Investigate different ways of filling these boxes with multiples of 5.

For example: 20% of 50 = 10, 80% of 25 = 20.

Can you find 30 different ways?

Unit 10.3 Percentage increase and decrease

In this unit you will:

- learn how to find a percentage increase or decrease.

Key words
Percentage increase
Percentage decrease
Unitary method

When a price goes up, it increases.
When a price goes down, it decreases.

Examples of **percentage increases** are:
8% more → 108%   10% extra → 110%   up by 25% → 125%

Examples of **percentage decreases** are:
5% less → 95%   35% off → 65%   reduced by 12% → 88%

To calculate the result of \$240 decreased by 15% we have to find 85% of the amount.
100% − 15% = 85%
Find 85% of \$240 by one of these methods:

Method 1 – Use fractions and cancel: $\frac{85}{100} \times \$\frac{240}{1} = \$204$

Method 2 – Use the **unitary method**: 1% of \$240 = \$2.40
85% of \$240 = 85 × \$2.40 = \$204

85% of \$240 is \$204 which is a decrease of \$36.

Worked examples

1. The original price of a large television is \$12 000. A sale offers a 20% discount on this price. What is the sale price?

2. A notebook costs \$12.00 plus VAT of 15%. What is the total cost of the notebook?

1. The original price of \$12 000 is 100%.
The 20% discount reduces the price to 80% of \$12 000.
80% of \$12 000 = $\frac{80}{100} \times \$\frac{12\,000}{1}$ – \$9 600.

2. 1% of \$12.00 is \$0.12.
So, 15% of \$12.00 is \$1.80.
\$12.00 + \$1.80 = \$13.80.
The total cost of the notebook is \$13.80.
Or:
The cost of the notebook is 100%. A 15% increase means 115% of \$12.00.
$\frac{115}{100} \times \$\frac{12.00}{1} = \$13.80$.

Exercise

1. Write the percentage increase or decrease represented by:

   a) 125% **b)** 150% **c)** 88% **d)** 101% **e)** 96.5%

2. Calculate the new prices after these percentage changes:

   a) Shirt $40, 10% increase
   b) Jacket $600, 25% decrease
   c) T-shirt $50, 20% increase
   d) Bicycle $300, 50% reduction
   e) Scooter $11 000, 30% extra
   f) Car $80 000, 15% off

3. Copy and complete this table to show the new amounts:

Amount	Increase by 10%	Decrease by 20%
$240		
$350		
$160		
$18		
$42		

4. In the January sale, everything in a souvenir shop has 20% off.
   What will each item sell for?

5. The population of a country increased from 120 000 to 150 000 over a period of 15 years. What is the percentage increase?

6. On average, class attendance in Form 4 increased from 30 in March to 33 in April. There are 40 students registered in Form 4.

   a) Calculate the percentage increase in attendance from March to April.
   b) On average, what percentage of students was present in March?
   c) On average, what percentage of students was absent in April?

Investigation

Choose any five islands in the Caribbean. Find out the population of each island in 1980, 1990, and 2000. Calculate the percentage increase/decrease in the population over each decade.

Comment on your findings. You may answer questions such as these:

a) Give reasons for the percentage increase/decrease.
b) What are the disadvantages and advantages of the decrease/increase?
c) What measures have governments put in place to cope with the changes?

Unit 10.4 Calculations with money

In this unit you will:

- work out the unit price of items given the total cost
- work out the total cost of items given the unit price and quantities
- perform calculations of discounts in percentage and money terms
- perform calculations involving sales tax or VAT.

Key words
Unit price
Total cost
Discount
Sales tax
Value Added Tax (VAT)

When you go to a shop, the items have prices on them. The cost of a single item is called the **unit price**. For example, a jar of jam costs \$3.50, a kilogram packet of meat costs \$9.75.

When you add up the cost of all the goods you are buying, you are calculating the **total cost** of the goods. For example, a jar of jam plus a kilogram packet of meat costs \$3.50 + \$9.75 = \$13.25.

Remember that money amounts are decimal. The amount after the decimal point indicates cents.

Sometimes the price of an item is discounted. This means that the unit price is reduced. Discounts are normally given as percentages. For example, John sees a book costing \$15.00. The store-keeper offers a 10% **discount** for cash. This means that John pays \$15.00 − 10% of \$15.00 = \$13.50.

In many places, government tax is added to the cost of goods. This tax is called **sales tax** or **Value Added Tax (VAT)**. Tax is also calculated as a percentage. For example, in Trinidad and Tobago, VAT of 15% is added to the total cost of all goods. This means that a bill of \$45.00 will cost \$51.75 after tax.

Duty-free goods do not attract sales tax.

Worked examples

1. James buys 12 mangoes for \$30.00. Calculate the unit price of a mango.
2. Zara buys a loaf for \$6.85, a bottle of milk for \$9.95 and a tub of margarine for \$12.99. What is the total cost of her goods?
3. A pair of shoes costs \$165.00. A discount of 10% is given. How much will the customer pay for the shoes?
4. The total cost of food items in Mr Stan's shopping basket is \$625.00. VAT of 15% is added to all food purchases. How much is Mr Stan's total bill?

1. \$30.00 ÷ 12 = \$2.50. The unit price is \$2.50.
2. \$6.85 + \$9.95 + \$12.99 = \$29.79. The total cost is \$29.79.
3. $\frac{10}{100} \times \$165 = \$16.50$ Or $100\% - 10\% = 90\%$
   $\$165.00 - \$16.50 = \$148.50$ $\frac{90}{100} \times \$165.00 = \$148.50$
4. $\frac{15}{100} \times \$625 = \$93.75$ Or $100\% + 15\% = 115\%$
   $\$625.00 + \$93.75 = \$718.75$ $\frac{115}{100} \times \$625 = \$718.75$

Exercise

1 Find the total cost of the following items:

a) 9 pens at \$9.95 each and 6 pencils at \$1.15 each

b) 2 packets of notebook paper at \$13.70 each, 4 spiral notebooks at \$22.59 each and 12 exercise books at \$1.20 each

c) 7 paper folders at \$1.10 each, 3 files at \$15.90 each and a stapler costing \$23.60

2 Mrs Pierre's shopping bill is as follows:

a) Find the total cost of the items on the receipt.

b) VAT of 8% is added. How much does Mrs Pierre have to pay?

RECEIPT

12 cans of milk at \$9.95 each
2.5 kg of potatoes at \$2.10 a kg
4 packets of bread at \$6.85 each
1 tin of coffee \$15.00
3 bathroom cleaners at \$8.50 each

3 Sheila bought the following clothing items:

10 T-shirts at \$35.00 each
3 pairs of pants at \$85.00 each
2 dresses at \$175.00 each

a) Calculate the duty-free price of the items.

b) Sheila has to pay 15% VAT as she does not qualify for duty-free. How much did she pay?

4 The cost of a dress is \$120.00. It was sold for \$96.00. Calculate:

a) the discount offered in dollars

b) the percentage discount.

5 The total cost of the items in Quincy's shopping basket at Sara's is \$78.00.

a) How much does Quincy have to pay for the items?

b) Quincy gives the cashier \$100.00. How much change does he receive?

6 Francis buys 11 mangoes.

a) How much does Francis pay?

b) Would it have been cheaper if Francis had bought 12 mangoes? If so, by how much?

7 A store offers 25% off items with a green tag and 40% off items with a yellow tag. Find the sale prices of the following items:

a) 

b) 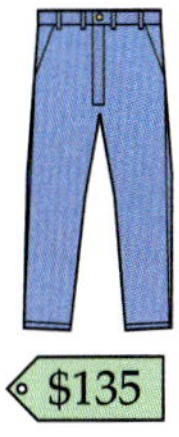

c) 

d) 

e) 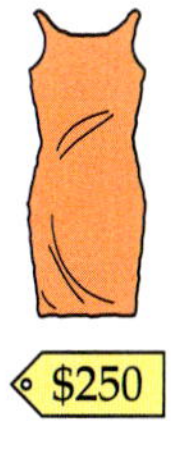

Unit 10.5 Hire purchase

In this unit you will:

- understand what is meant by hire purchase (HP)
- perform calculations to compare prices, work out interest and find total costs related to hire purchase agreements.

Key words
Credit
Deposit
Balance
Instalments
Hire purchase
Interest

Sometimes people want to buy something but they do not have enough money to pay for it immediately. In these cases, the shopper can buy on **credit**. In other words, the shopper pays part of the price as a **deposit** and agrees to pay the rest of the amount owed, or the **balance**, in **instalments** over a fixed period of time.

CASH **$299**
Deposit $45 Six payments of $50.80

A common form of buying on credit is called **hire purchase**. In this system, the shopper pays a deposit and signs an agreement to pay fixed monthly amounts till the goods are paid for. The shopper can take the goods home but they technically belong to the shop until they are paid for in full.

Hire purchase costs more than when you pay cash because the shop charges **interest** on the outstanding balance.

To calculate the hire purchase price, add the deposit to the total of the monthly payments. In the example above the hire purchase price is:

$$\$45 + (6 \times \$50.80) = \$45 + \$304.80 = \$349.80$$

The interest can be calculated using a formula:

$$\text{Hire purchase price} - \text{Cash price} = \text{Interest paid}$$

Using the example above: $\$349.80 - \$299.00 = \$50.80$

The customer paid $50.80 interest on hire purchase.
This means it would be $50.80 cheaper to pay cash for the bicycle.

Worked examples

The cost price of a mantelpiece is $1 500. A shopper makes a deposit of 15% and pays 12 monthly instalments of $180. Find:

a. the hire purchase price **b.** the interest earned.

a. $\frac{15}{100} \times \$1\,500 = \$225$

$12 \times \$180 = \$2\,160$

Hire purchase price $= \$2\,160 + \$225$

$= \$2\,385$

b. Interest $= \$2\,385 - \$1\,500$

$= \$885$

Exercise

1. The cash price of a table is $795 and the hire purchase price is $900. Find the interest charged on hire purchase.

2. Tonya buys a television on hire purchase. She pays a deposit of $150 and then pays 24 monthly instalments of $116. Calculate the hire purchase price of the television.

3. Matthew buys a settee on hire purchase. He pays a deposit of $400 plus 18 monthly instalments of $300. Find:

   a) the hire purchase price
   b) the interest paid.

4. The cash price of a bed is $2 195. Luke bought the bed on hire purchase. He paid a deposit of $250 and 24 equal monthly instalments. The hire purchase price was $4 500. Calculate:

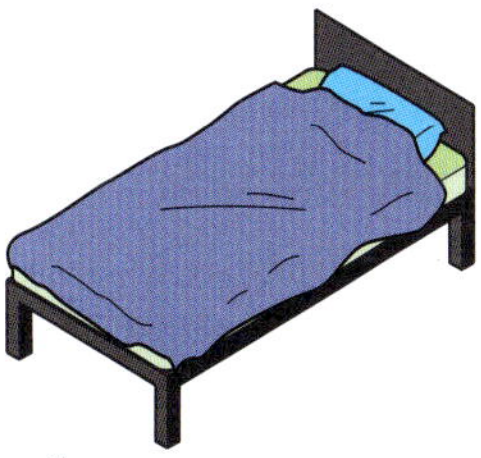

   a) how much interest he paid **b)** his monthly instalment amount.

5. The cash price of a stove is $4 500. The stove was bought on hire purchase. There was no deposit and 24 monthly instalments of $250 were paid. Calculate:

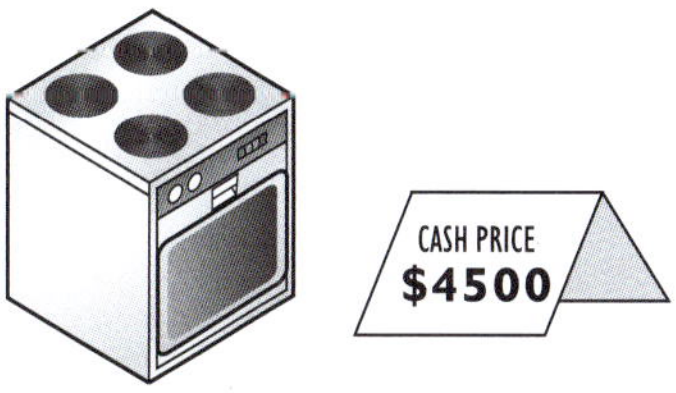

   a) the hire purchase price
   b) the interest charged
   c) the percentage interest charged on HP.

6. The cash price of a fridge is $8 000. Calculate:

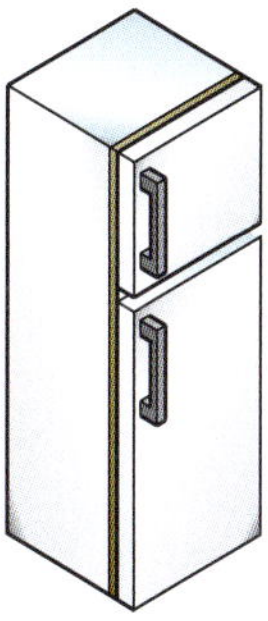

Pay cash $8 000 or 15% deposit + 36 monthly payments of $250

   a) the deposit in dollars
   b) the hire purchase price
   c) the amount paid in interest.

7. The cash price of a computer is $3 800. It was sold on hire purchase at an interest rate of 60%. Calculate:

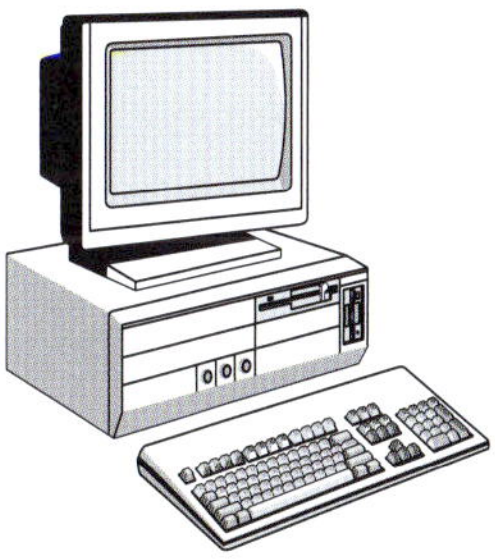

   a) the hire purchase price
   b) the interest in dollars.
   c) If no deposit was paid, calculate the monthly instalments if equal payments were made over 16 months.

8. A dressing table was bought on hire purchase. A 20% deposit was paid plus 20 monthly instalments of $360. The hire purchase price was $8 000. Calculate:

   a) the deposit in dollars **b)** the cash price
   c) the interest charged on hire purchase as a percentage.

Unit 10.6 Profit and loss

Key words
Profit
Loss

In this unit you will:

- learn what is meant by profit and loss
- calculate profit and loss in money and as a percentage
- work out cost price and selling price given information about profit and loss.

Many people make a living out of buying and selling. In such businesses, people buy goods or make them from raw materials and then sell them hoping to make some money in the process. The money paid to buy goods, or materials to make goods, is called the cost. The money made from selling the goods is called income or revenue.

When a business sells goods for more than it cost to buy or make them, it earns some money. This money is called a **profit**.

When the business is forced to sell goods for less than it cost to buy or make them, it suffers a **loss**.

For example, Rosalyn buys a small loom for $450 and a stock of wool for $150. She weaves 90 place mats and sells them all for $7.50 each.

Her income was: 90 × $7.50 = $675
Her costs were: $450 + $150 = $600
She made $75 profit.

Rosalyn's neighbour also buys a small loom and a stock of wool for the same prices. She decides to sell her place mats for $6.00 each instead of $7.50.

Her income was: 90 × $6.00 = $540
Her costs were also $600
This means she made a loss of $60.00.

Profit = Selling price − Cost price

$$\% \text{ Profit} = \frac{\text{Profit}}{\text{Cost price}} \times 100$$

Loss = Cost price − Selling price

$$\% \text{ Loss} = \frac{\text{Loss}}{\text{Cost price}} \times 100$$

Worked examples

1. The cost price of a chair is $200. It was sold for $250. Calculate the profit and the percentage profit.
2. A table bought for $600 was sold at a loss of 10%. Find the selling price of the table.
3. A house was sold for $150 000 with a profit of 25%. Find the cost price of the house.

1. Profit = $250 − $200 = $50   % profit = $\frac{50}{200} \times 100 = 25\%$
2. $\frac{10}{100} \times \$600 = \$60$ The selling price was $600 − $60 = $540.
3. 100% + 25% = 125%

   125% represents $150 000

   1% represents $\dfrac{\$150\,000}{125}$

   Cost price: 100% represents $\dfrac{\$150\,000}{125} \times 100 = \$120\,000$

   The cost price was $120 000.

Exercise

1. Copy and complete the following table:

Cost price	Selling price	Profit	Loss
$300	$320		
	$100	$60	
$25			$5
$1 350	$1 275		
	$480		$65
$798		$42	
	$200	$0	
$2 500	$3 100		
$5 120	$4 765		
	$79		$34

2. The cost price of a fan is $120. It was sold for $140. Find the profit and the percentage profit.

3. The cost price of a printer is $450. 25 printers were sold to a school for a total of $10 000.
   a) Was a profit or loss made? **b)** How much was the profit/loss?

4. A settee bought for $1500 was sold at a profit of $12\frac{1}{2}$ %. Find:
   a) the profit **b)** the selling price of the settee.

5. A car was bought for $96 000. Eight years later it was sold for $89 000. Calculate:
   a) the loss made **b)** the percentage loss.

6. A computer was sold for $2 400 with a loss of 60%. What was the cost price of the computer?

7. A stereo was sold for $3 600 with a profit of 20%. Calculate the cost price of the stereo.

8. Harry and Paul each bought a bicycle at different stores. Harry sold his bicycle for $900 and made a 50% profit. Paul sold his for $1 110 and made a 48% profit.
   a) Calculate the cost price of each bicycle.
   b) Which was the more expensive bicycle?
   c) Who made the greater profit?

Investigation

Obviously people do not want to sell their goods at a loss. Investigate why shopkeepers might choose to sell certain goods below cost. Make a list of reasons that you find out.

Unit 10.7 Simple interest

In this unit you will:

- learn how to calculate simple interest
- apply formulae to solve problems related to simple interest.

Key words
Simple interest
Principal
Rate
Time
Per annum

One of the ways in which banks make money is by charging interest on loans given to the customers. One of the types of interest is **simple interest**.

$$I = \frac{P \times R \times T}{100}$$

Where I = simple interest
P = **principal**, i.e. the amount of the loan
R = **rate**, i.e. rate of interest as a percentage
T = **time**, the number of years given for repaying the loan.

The rate of interest is usually given as the interest rate **per annum**, i.e. for a period of a year.

Worked examples

1. Find the simple interest on a loan of $500 for 2 years at 6% per annum.
2. How much is paid to the bank after 4 years?

Note: Most banks calculate interest at a compounded rate.

1. $\frac{500 \times 2 \times 6}{100} = \$60$

   The interest was $60.

2. Total interest = $60 × 2 = $120

   Amount paid to bank = $500 + $120
   = $620

Exercise

1. Calculate the simple interest due on a loan of:

   a) $1 200 for 3 years at 10% p.a.
   b) $600 for 2 years at 6% p.a.
   c) $1 000 for 2 years at $12\frac{1}{2}$% p.a.
   d) $3 200 for 4 years at 12% p.a.
   e) $18 000 for 5 years at 9% p.a.
   f) $20 000 for 3 years at 6.5% p.a.
   g) $400 for 2 years at 8% p.a.
   h) $9 000 for 6 years at 13% p.a.
   i) $120 000 for 20 years at 8.5% p.a.
   j) $5 000 for $2\frac{1}{2}$ years at 6% p.a.

2. Calculate the total amount paid to the bank by the end of the specified time period in each of the above.

Case study - Mrs Jenny Boyce (known as 'Ma Boy')

I live in a village called Little Town and I am known as the village's caterer. I cater for weddings, parties, lunches and dinners … as long as you need someone to bake or cook for a gathering, I am the one that you should call on.

I learnt to cook and bake from the tender age of 8. At that time, I was able to cook my mother's favourite dish – green fig and saltfish with a cucumber salad. I was originally taught to cook by my grandmother, who was at that time a very popular cook. Over the years, I taught myself by making use of various recipes from my numerous cookbooks.

Although I like cooking, I dislike calculating the quantities of items that I need to buy based on the recipe that I use and the number of people that I cater for. I also have quite a bit of difficulty figuring out how much I should charge my clients if I have to make a reasonable profit. I can understand why I have difficulty with these calculations; mathematics was my least favourite subject at school – I seldom paid attention. Little did I know that I would have to use mathematics in this business of catering. However, as luck would have it, my niece loves mathematics and gives me a hand with problem-solving from time to time. So, most times I do make a decent profit and can earn my living.

This morning, I received an order from a restaurant for 20 coconut cakes. Here are the ingredients needed to make 1 coconut cake:

$2\frac{1}{2}$ cups flour
$\frac{1}{2}$ cup milk
$1\frac{1}{2}$ teaspoons baking powder
$1\frac{1}{2}$ cups grated coconut
2 eggs
$\frac{1}{2}$ teaspoon vanilla essence
1 cup granulated sugar
$\frac{1}{2}$ cup shortening

Activity

Help Ma Boy by working out:

a) The quantity of each item needed to make 20 coconut cakes

b) The amount to be spent on shopping for these items.

Note: There is VAT of 15% and she does not need to buy baking powder or vanilla essence.

c) The minimum amount that she should charge the restaurant owner for one cake in order to make a profit.

Note: She spends about 5 hours baking the cakes and she charges $40.00 per hour for her time.

Make use of the following information:

4 cups of flour	**= 1 lb flour**
2 cups granulated sugar	**= 1 lb granulated sugar**
16 cups milk	**= 1 gallon milk**
4 cups grated coconut	**= 1 grated coconut**
2 cups shortening/ margarine	**= 1 lb margarine**

1 lb flour costs $1.50
1 dozen eggs cost $6.00
1 gallon milk costs $20.00
1 coconut costs $1.00
1 lb margarine costs $8.00
1 lb granulated sugar costs $2.00

Looking back

1. A teacher had 120 sweets in a bag. She shared 90% of the sweets equally among her students. If each student received 3 sweets, how many students were in the class?

2. There were 2000 registered voters in a certain constituency. 42% voted for Party A, 900 voters voted for Party B and the rest of the voters did not vote.
   a) What percentage of voters voted for Party B?
   b) How many voters did not vote?

3. 12% of the students in Form 4 have the flu. 44 of the students do not have the flu. How many students are registered in Form 4?

4. The table opposite shows how Mr Charles uses his monthly earnings of $2 500.
   a) How much money does Mr Charles save each month?
   b) What fraction of his earnings is spent on utilities?
   c) How much does he spend on food?

Utilities	32%
Savings	40%
Loan	10%
Food	?

5. The cost of the items in Mrs Stanford's shopping basket is as follows:

   3 shower curtains @ $48.50 each
   120 sweets (15 sweets for $1.00)
   4 tins of air freshener @ $15.00 each
   1 coffee maker for $75.00
   3 packets of flour @ $2.50 each

   a) Find the total cost of the items in Mrs Stanford's shopping basket.
   b) VAT of 15% was added. How much would Mrs Stanford have to pay?
   c) How much change would Mrs Stanford receive from $350.00?
   d) Calculate the price of each item in a '15% off' sale.

6. 12 oranges cost $1.50. Find the cost of
   a) 8 oranges b) 18 oranges c) 42 oranges

7. The cash price of a microwave is $2 400. If bought on hire purchase, a deposit of 20% is required plus 12 monthly instalments of $275. Calculate:
   a) the hire purchase price
   b) the interest paid in dollars.

8. A refrigerator cost $7 500. It was later sold for $5 400. Calculate the loss and the percentage loss made.

9. A dining table sold for $2 600 with a profit of 30%. What was the cost price of the dining table?

10. Calculate the simple interest on a loan of:
    a) $4 000 for 2 years at 9.5% per annum b) $16 000 for 5 years at 12% per annum

Project

People are paid in different ways for the work they do. Some of these ways are listed in the table below. Other people work for themselves – we say they are self-employed. These people may charge a fee for their services or sell goods for a profit.

Type of payment	What this means	Examples
Salary	Salary is the amount paid for a whole year's work. It is normally paid monthly. Salaried workers are entitled to paid holidays, paid sick leave and other benefits. Salaried workers are not normally paid for overtime.	Teacher, banker
Wages	Wages are the amount paid on an hourly rate for hours worked in a period (normally a week). Workers who earn wages are usually entitled to paid holidays, paid sick leave and overtime if they work on public holidays. Overtime is normally paid at a higher hourly rate than normal work.	Builder, factory worker
Casual	A fixed hourly rate for hours worked. This is not a permanent position and no benefits are given.	Waiters in fast food places, cleaners
Commission	Commission is a percentage of the value of goods sold. Benefits will depend on an agreement between employer and worker.	Salespeople, Insurance brokers

1. Find the employment section from a newspaper. These are the pages where jobs are offered. Bring these to class.
2. For each of the types of payment mentioned above, try to find two examples of jobs that are paid in that way. Cut out the advertisements and stick them into your book.
3. Label the advertisements to show how much is paid and by what method it is paid.
4. Detail any extra payments that are offered, for example overtime, holiday pay or bonuses.
5. Work in groups of four. Discuss the advantages and disadvantages of each method of payment.
6. Find out what deductions are made from people's earnings. You can ask people in the community to help you find this information.
7. Write a summary of what you have learned and be prepared to report your group's conclusions to the class.

R3 Revise and consolidate

1. Draw flow diagrams to show how to work backwards to find the value of the following variables.

   a) $x + 4 = 13$ b) $\frac{h}{8} = 7$ c) $4t = 80$

   d) $m - 19 = 12$ e) $2x + 5 = 23$

   f) Check your results to be sure that the equations balance.

2. Write the instructions for each of the following:

   a) $7x + 5 = 9$ b) $y - 4 = 12$ c) $8a \div 6 = 2$

   d) $25 - p = 4$ e) $c - 9 = 4c$

   For example: $3b - 9 = 7$; three times a number, subtract 9 and the result is 7.

3. Five bags of flour have a mass of 20 kg.

   a) Write an equation using x to stand in for a bag of flour.

   b) Solve the equation to find the mass of each bag of flour.

4. Copy and complete this table:

	Cost price	Selling price	Profit or loss in $	% profit or loss
a	$100	$120		
b	$1.00	$1.50		
c	$55	$66		
d	$160	$150		
e	$50		$6 profit	
f		$1.40	20c loss	
g	$10			12.5% profit
h	$0.50			40% loss
i		$13		30% profit
j		$1.60		4% loss

5. A family home is insured for $28 500. The householder has to pay 8% of the value per annum as premiums.

   a) Calculate how much the householder pays in total each year.

   b) How much will this be per month?

Investigation

How many angles are in the diagram?

Use a term or an expression to name each angle. Did you have the angles $4a$ and $2a + 3a$ in your list?

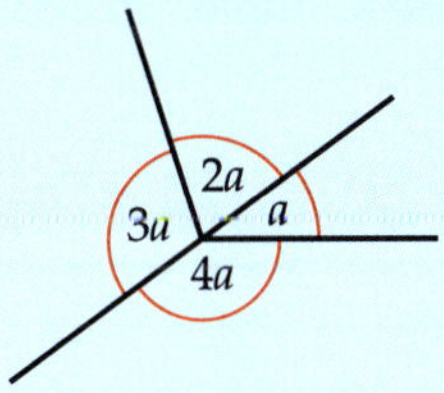

Investigation

In a test, Fay got twice Roy's marks; Jay got 40 more marks than Roy. The total marks for the three children were 160. What did each child score on the test?

Investigation

Input any five numbers in the following flow chart. What do you notice about the pattern of the output?

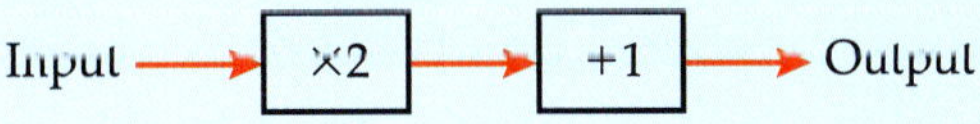

Investigation

Ma Belle sees a sign advertising 'up to 50% off all goods' in the window of her local clothing store. She goes in and buys a dress for $595, a pair of shoes for $375 and a hat for $234. The total cost of the goods is $1204.00. Ma Belle expects to pay $602.00, but when the assistant rings it up, it comes to $698.32. What do you think happened?

Explain what is meant by:

- bulk discount
- buy one get one free
- below cost
- no duty
- service charge

Investigation

Work in groups of four. Find a department store in your town which offers hire purchase agreements.

Choose four items and work out:

- the hire purchase price of such items
- the percentage charged in interest on each item.

Make a list of the advantages and disadvantages of buying goods on hire purchase.

Chapter 11

Lines, angles and shapes

This chapter deals with lines, line segments, rays and angles. You will learn to recognise, name and draw different types of line segments and lines. You will then name, describe and measure angles. Angles will also be classified by size and you will use the relationship between angles to calculate missing angles.

Key words
Point
Line
Line segment
Ray

Unit 11.1 Lines, line segments and rays

In this unit you will:

- differentiate between lines, line segments and rays
- measure and draw line segments of different lengths.

In mathematics, a **point** is a position. Points have no dimensions. In other words, a point has no length, width or height. Points can be used to show the position of towns, buildings or other geographical attributes.

In the diagrams below A, B, M and N show different ways of representing points in geometrical drawings.

A **line segment** is the distance between two points. It represents length but not width. The letters at either end are used to name the line segment.

The distance from point A to B is the line segment AB. This line segment can also be called BA.

Line segments can be straight as in AB or curved as in MN.

In mathematics, a **line** has no end points; it goes to infinity in both directions. Arrows at the end of a line segment show that it continues indefinitely. Any two points on the line can be used to name the line.

A ray has one end point but continues to infinity in the other direction. BA is a ray. It starts at B and continues in the direction of A.

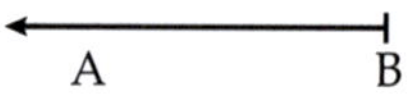

Mathematical meanings

In mathematics it's important to be clear about what we mean. We use special mathematical vocabulary to help us be clear in our meanings. However, we also use some of these terms in different ways in our everyday lives. Think about how we use the word 'point', for example:

The word 'line' is often used to describe line segments. For example, we say the shortest distance between two points is a straight line. In mathematics we know this is actually a line segment.

Investigation

Choose five mathematics words from the box below. Draw up a table comparing their meaning in mathematics and in everyday life. One example has been done for you:

Term	Mathematical meaning	Everyday meaning
Line	An unending line segment	Row, queue, boundary, ancestry

approximate average base cone cube degree
digit difference edge equal factor intersection
length negative net odd plane sequence
solid sphere sum total unit zero

Drawing points accurately

You need to use a sharp pencil for maximum accuracy. Look at the drawings of line segments on page 108. Note the position of the centres where the cross or stroke crosses a line or line segment – the centres represent the points.

Measuring and drawing line segments

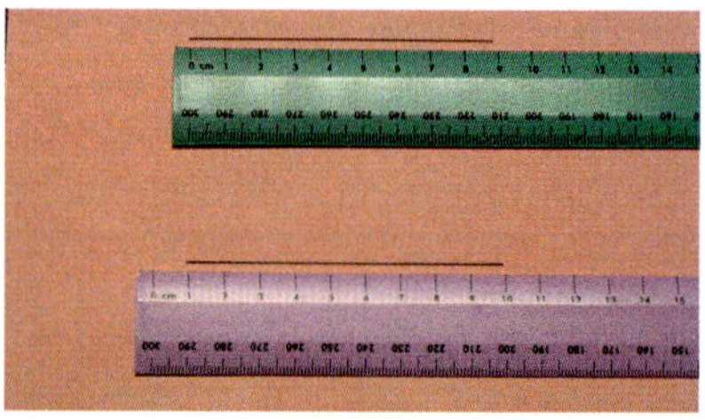

When you measure a line segment, you should try to place the zero marking of your ruler against one end point. If not, you have to take care to count off the units to get an accurate measurement. You can see two methods of measuring a 9 centimetre line segment in the photograph.

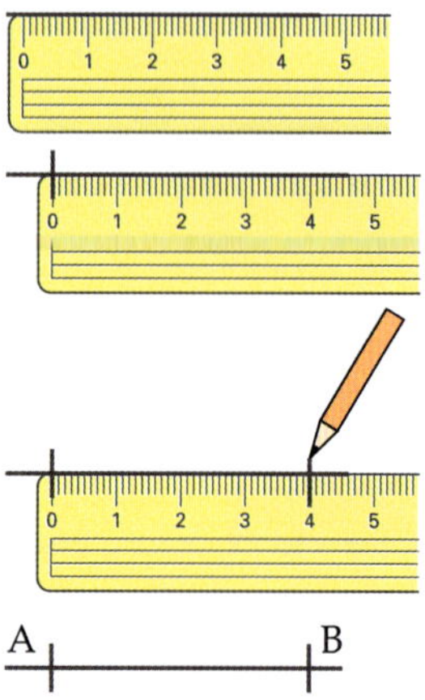

To draw a line segment of a given length:

- Draw a straight line with your ruler.
- Mark off one end point by making a cross or a stroke.
- Place your ruler on the line segment and measure the required length.
- Mark the second end point the correct distance from the first.
- Label the end points.

Exercise

1. State whether each sketch shows a line, a line segment or a ray.

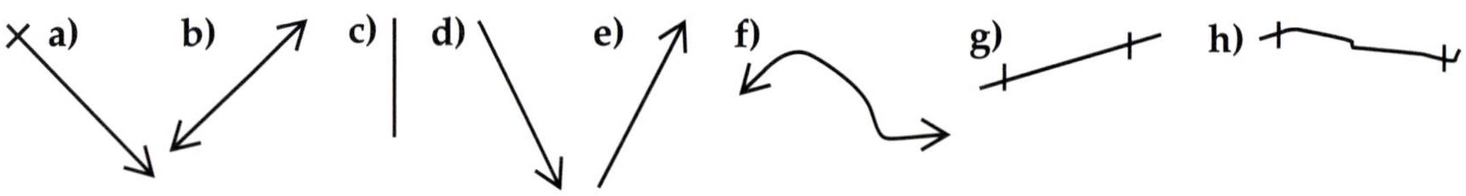

2. Make a sketch to show the following situations. Indicate whether you have drawn a line, a line segment or a ray.
   - **a)** Margaret stands at her classroom door and looks west.
   - **b)** The distance between Bright City and Moonlight City.
   - **c)** The direction east to west.

3. Measure the following line segments in centimetres.
   - **a)**
   - **b)**
   - **c)**
   - **d)**
   - **e)**

4. Draw line segments of the following lengths:

   a) 5 cm **b)** 35 mm **c)** 6.5 cm **d)** 67 mm

5. Work in pairs. Draw and label five line segments of lengths supplied by your partner. Check each other's work for accuracy.

Key words
Intersect
Angle
Arms
Vertex

Unit 11.2 Naming angles

In this unit you will:

- identify and describe angles.

When lines, line segments, rays or planes meet or cross we say they **intersect**.
An **angle** is the measure of turn or rotation between rays, line segments, lines or planes. The angles in the diagrams below are shown by curved lines with arrows.
Angles are measured in degrees.

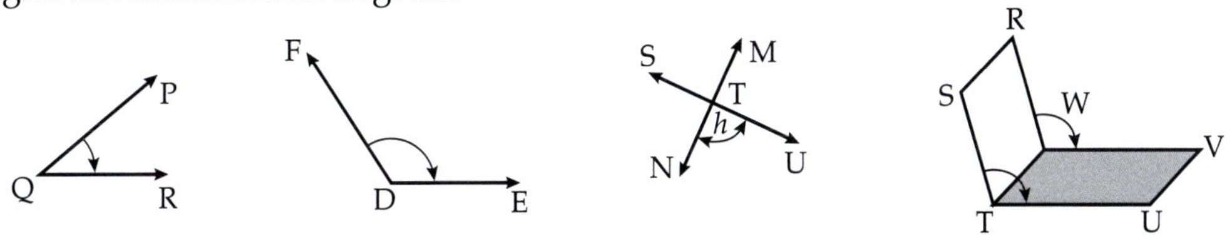

Look at the first diagram.
PQ and RQ are the **arms** of the angle.
Q is the **vertex** where the line segments intersect.

We can name angles in different ways. The angles shown above are:

- $\hat{Q}$, ∠RQP or ∠PQR
- $\hat{D}$, ∠EDF or ∠FDE
- ∠NTU or ∠UTN or *h*
- ∠STU and ∠RWV

Note that the angle symbol can be placed above the letter representing the vertex as in $\hat{Q}$ or before the three letters representing the angle as in ∠ABC. When three letters are used, the middle letter is the one at the vertex. When a lower case letter is used, as in *h*, no angle symbol is used.

Exercise

1. Use letters to name the arms of each angle and then name each angle.

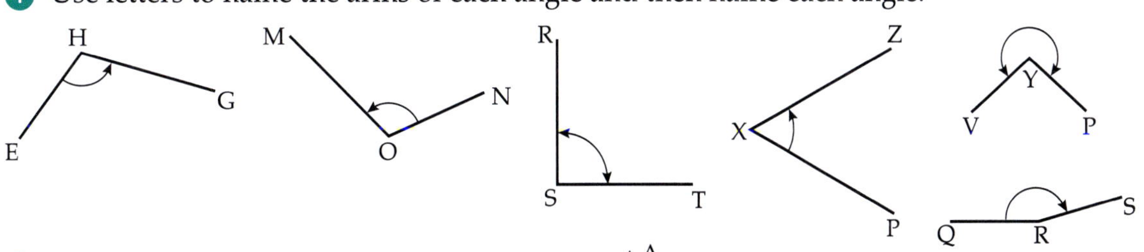

2. Name all the angles you can find in the following diagram.

3. Draw a labelled diagram of your own. Challenge a partner to name all the angles you have formed.

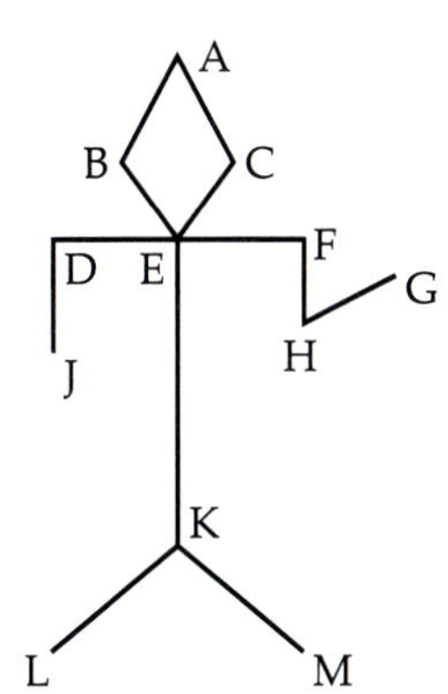

Unit 11.3 Angle sizes

In this unit you will:

- use size to name angles by type
- estimate and measure the size of angles.

Key words
Revolution
Straight angle
Right angle
Perpendicular
Acute angle
Obtuse angle
Reflex angle
Protractor

Angles are measured in degrees from 0° to 360°.
One full turn, or 360°, is called a **revolution**.
The hands of a clock turn through an angle of 360°.

A half turn, or 180°, is called a **straight angle**.
A quarter turn, or 90°, is called a **right angle**. Rays, lines, line segments or planes that form right angles are **perpendicular** to each other.

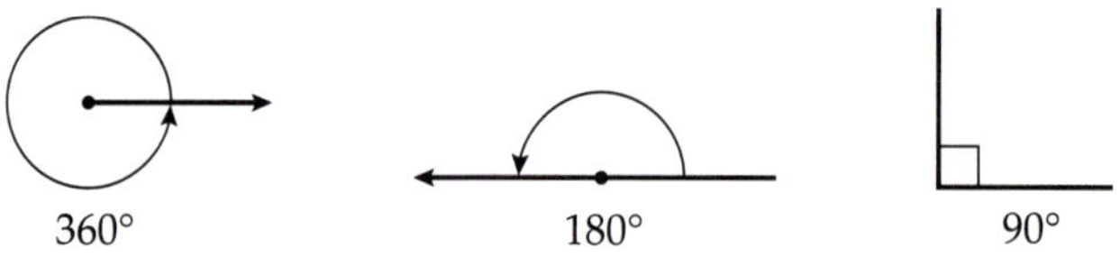

Angles which are less than 90° are called **acute angles**.
Angles which are greater than 90° but less than 180° are called **obtuse angles**.
Angles which are greater than 180° but less than 360° are called **reflex angles**.

Investigation

1. Look around you to find lines or surfaces that are perpendicular.
   a) Find the right angle in each example you find.
   b) Discuss in groups what problems might arise in a building if the walls were not perpendicular to the floor.

2. Try to measure the height of your classmates.
   Why is it important for them to stand perpendicular to the floor to do this?

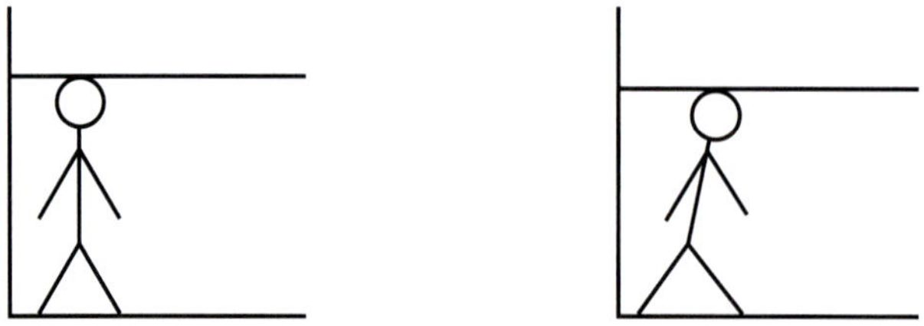

3. Do some research of your own to find the relationship between one revolution of the Earth around the Sun and the number of degrees in a full turn.

Measuring angles

We use a **protractor** to measure angles. The photograph shows you how to measure an angle with a protractor. The zero line of the protractor must be on one of the arms of the angle to start. You'll notice that the direction in which you measure does not matter as long as you read the angle from the correct scale. You'll also notice that the length of the arms does not affect the size of an angle.

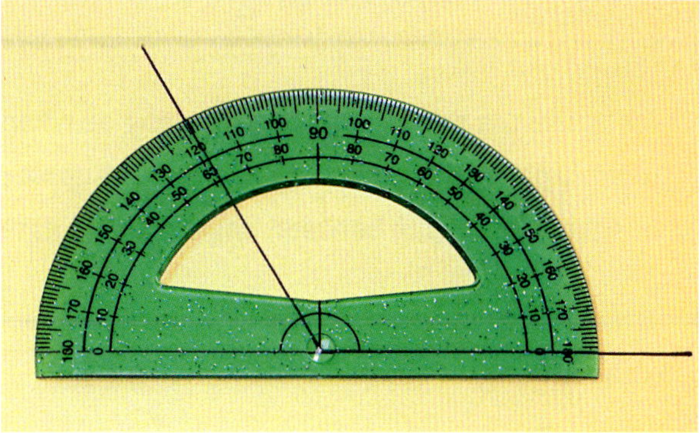

Exercise

1. Work in pairs. Draw ten angles of different sizes. Exchange angles and measure each other's angles with a protractor. Check your partner's work for accuracy.

2. Find ten angles in the classroom. Classify them as acute, right, obtuse or reflex.

3. **a)** Group the following angles in a table. Use the headings acute, obtuse and reflex.

18°	115°	70°	36°	160°	215°	93°
185°	86°	5°	340°	295°	155°	143°

   b) Draw $A\hat{B}C$ the smallest acute angle and $O\hat{B}E$ the largest obtuse angle in the table.

4. **a)** Estimate the size of the following angles and write down your estimates.

   b) Use a protractor to measure each angle. Write the measurements next to the estimates. You may have to extend one of the arms.

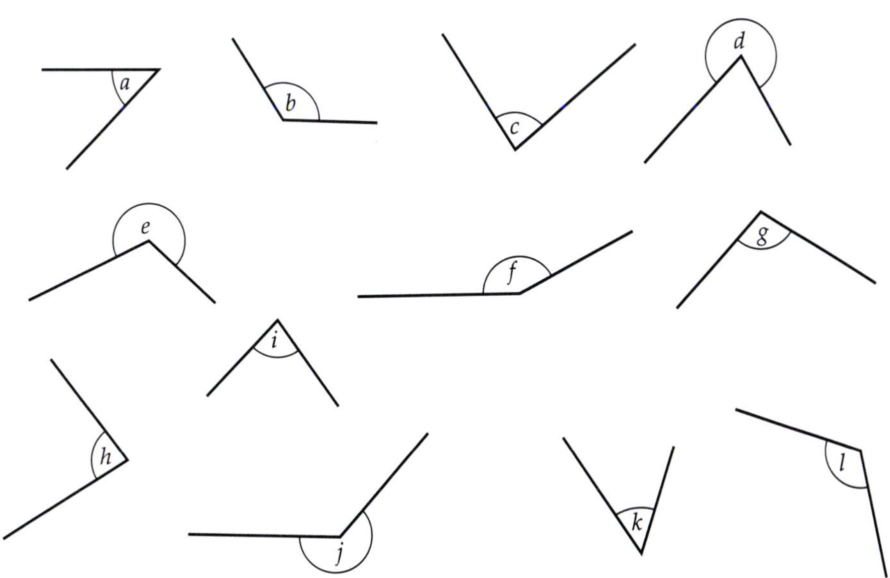

Unit 11.4 Related angles

In this unit you will:

- recognise the relationship between angles when lines intersect
- identify pairs of angles that are equal or supplementary
- use the relationship between angles to calculate missing angles.

Key words
Adjacent
Supplementary
Complementary
Vertically opposite

Adjacent angles share a common arm.

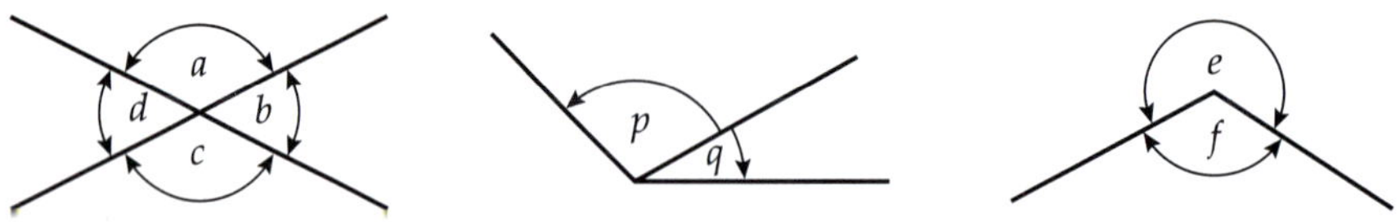

a and *b*, *p* and *q*, and *e* and *f* are adjacent angles.

Angles which meet at a point on a straight line add up to 180°. These are called **supplementary** angles. The angles *a* and *b* above are supplementary so $a + b = 180°$.
Angles which meet at a point add up to 360°. In this example $a + b + c + d = 360°$.

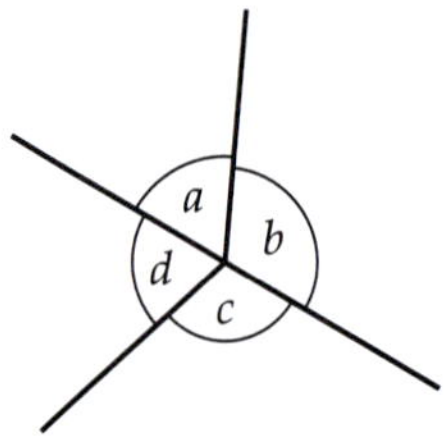

Angles which add up to 90° are called **complementary** angles.

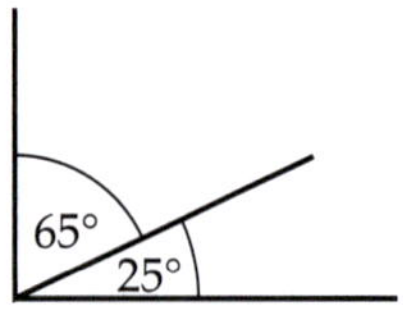

If two lines intersect, four angles are formed. The two angles in a pair of **vertically opposite** angles are equal. Vertically opposite angles meet at a common vertex.
In this example $a = c$ and $b = d$ because they are vertically opposite.

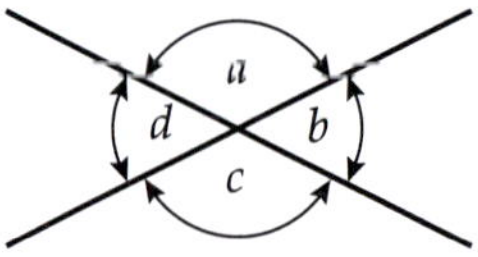

Investigation

Draw two lines which intersect in an *X* shape.

Name the angles formed.

Use folding to compare the sizes of the adjacent and opposite angles.

Will this work with any two intersecting lines?

Worked example

Find the size of the lettered angles. Give reasons for your answers.

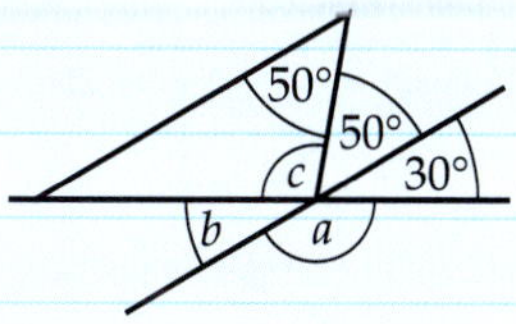

a is 150° because angles on a straight line add up to 180°.

b is 30° because it is vertically opposite to an angle of 30°.

c is 100° because all the angles at a point add up to 360°.

Exercise

1 Use what you have learned to classify the following pairs of angles. Write the name of each pair.

a)

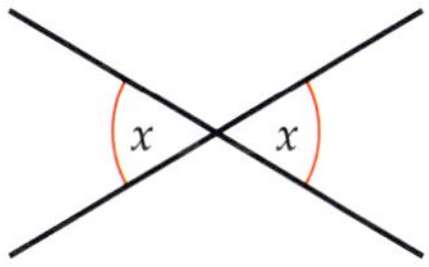

b)

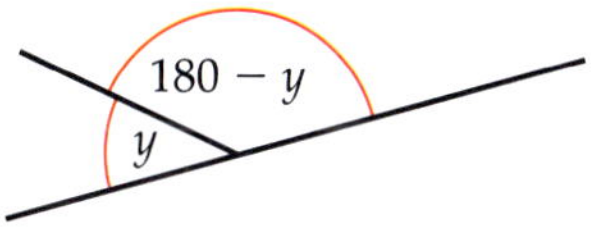

c)

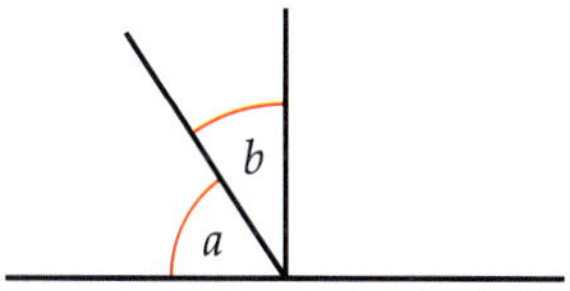

d)

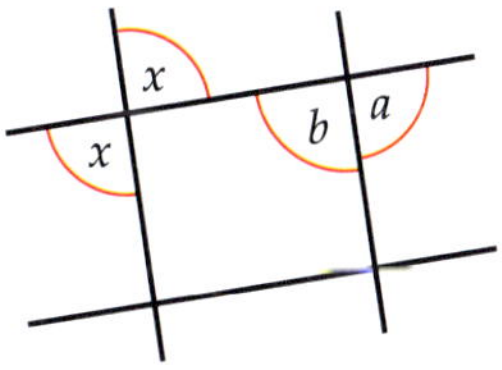

e)

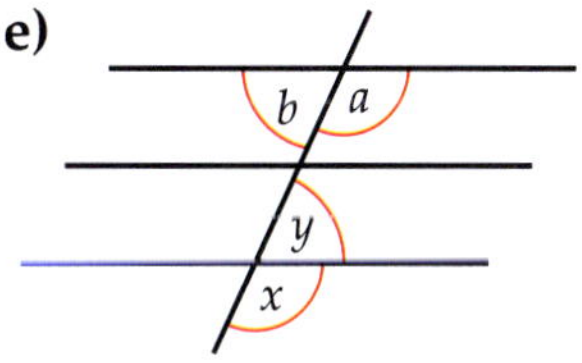

f)

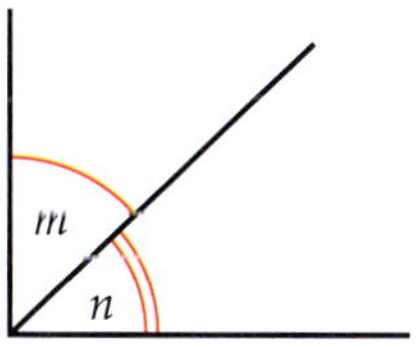

2 Name:

a) one pair of vertically opposite angles

b) one pair of adjacent angles

c) one pair of supplementary angles

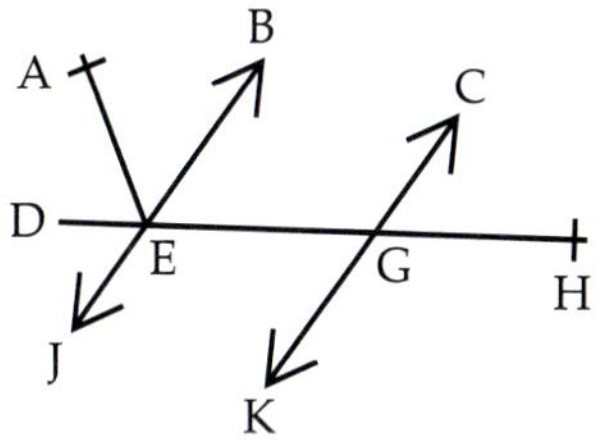

3 Calculate the unknown angles in the following diagrams. Give reasons for your answers.

a)

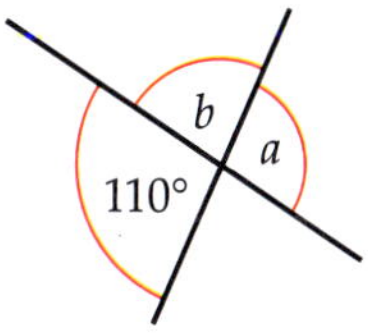

b)

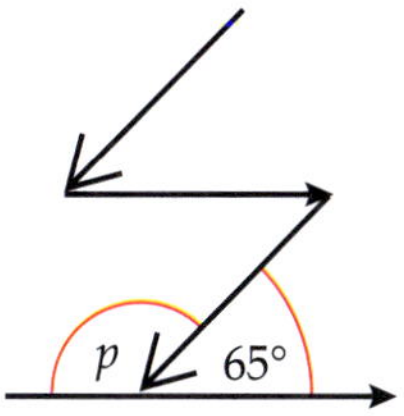

c)

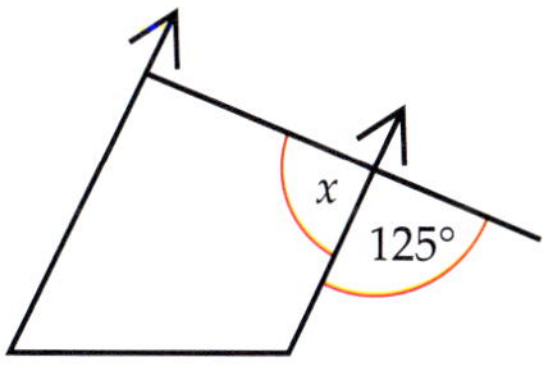

d)

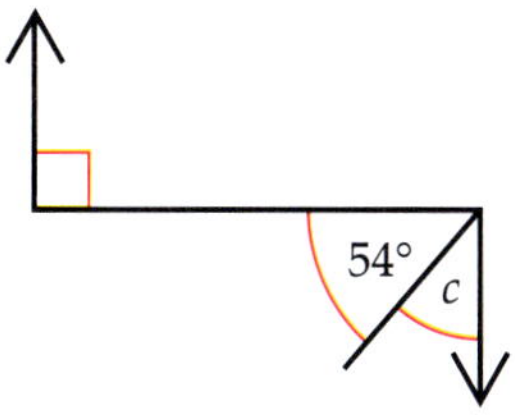

e)

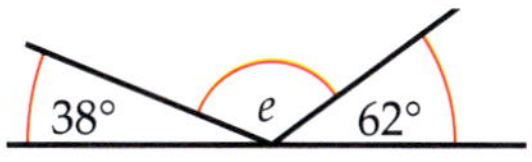

4 Calculate the unknown angles. Give reasons.

a)

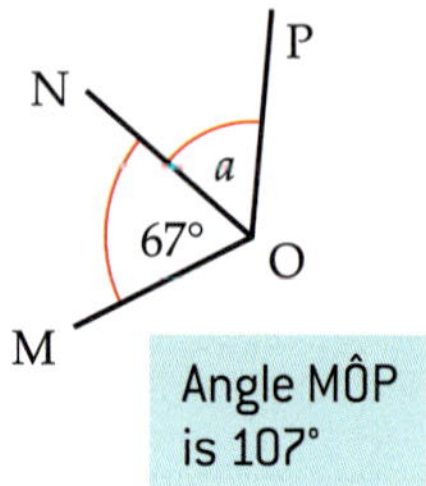

b)

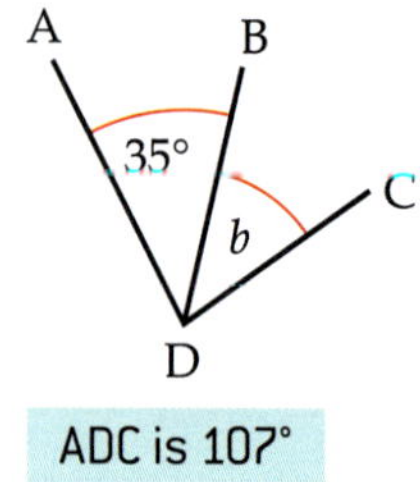

c)

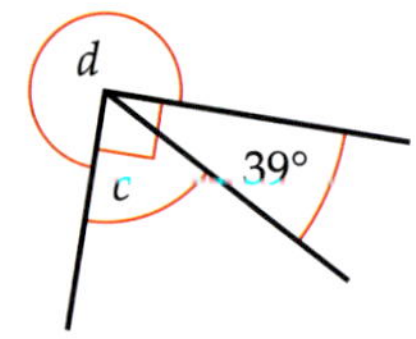

d)

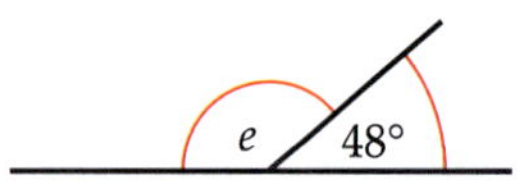

e)

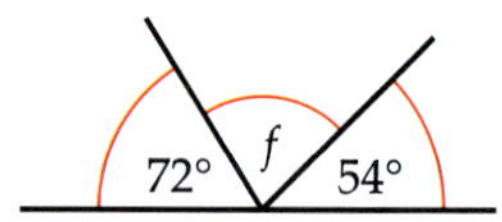

f)

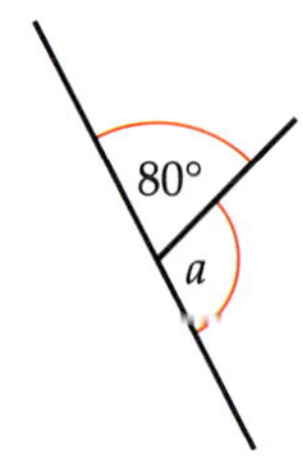

g)

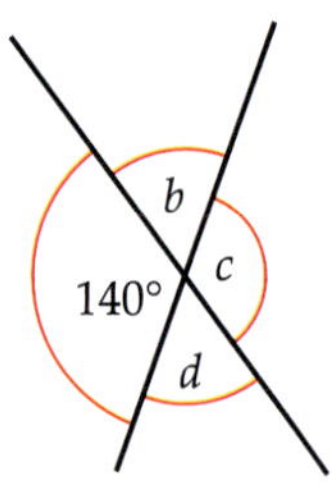

h)

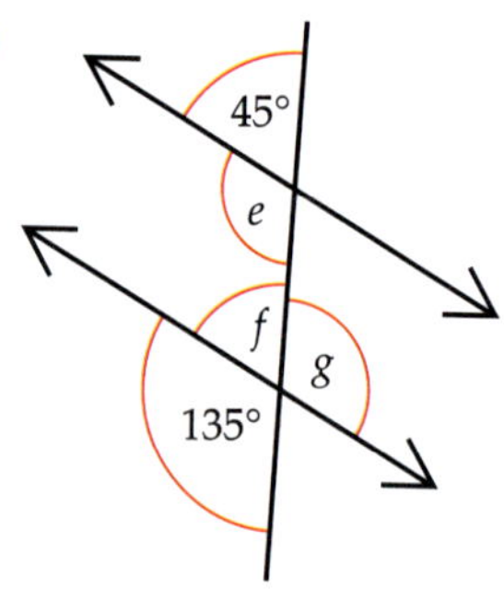

5 The angles of a rectangle are all 90°. The opposite sides are parallel. Calculate the value of p and q. Give a reason for each answer.

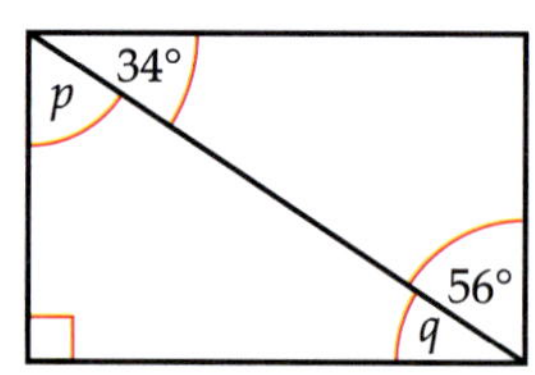

6 Find the size of x, y and z in this diagram.

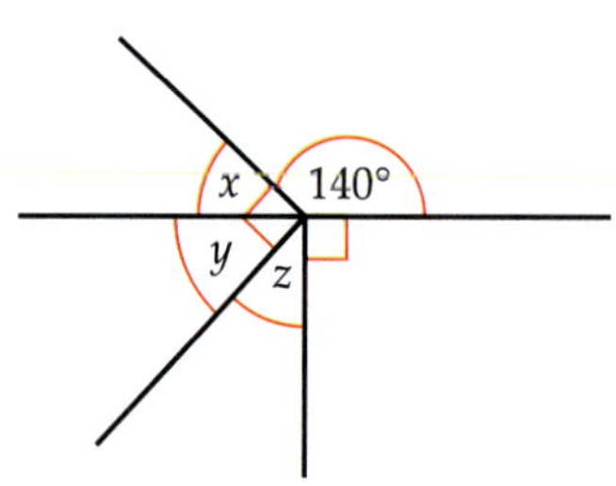

7 Use a ruler and protractor to draw:

a) supplementary angles x and y, $y = 120°$

b) complementary angles m and n, $m = 20°$

c) vertically opposite angles p and q, $q = 75°$

Looking back

1 In the following diagram:

a) name three points.

b) how many angles are at A? Name them.

c) name a line segment.

d) name a ray.

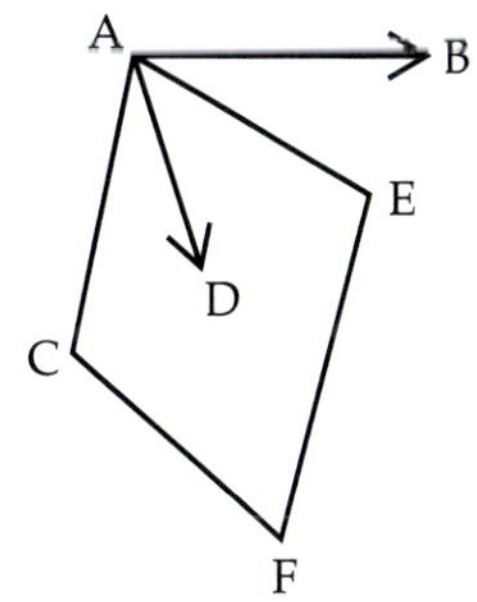

2 Draw the following angles and state whether they are acute, obtuse, straight, right-angled or reflex.

a) 109° **b)** 100° **c)** 245° **d)** 180° **e)** 15° **f)** 90°

3 Estimate then measure the following angles:

a)

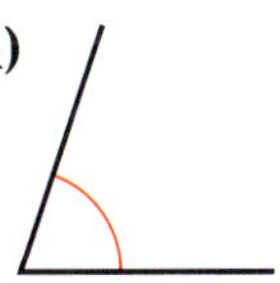

b)

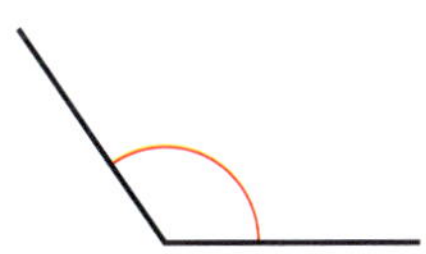

c)

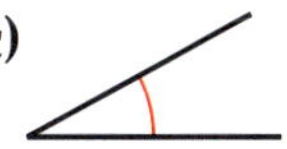

d)

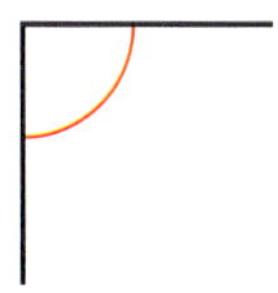

4 Calculate the sizes of the following angles. Give reasons.

a)

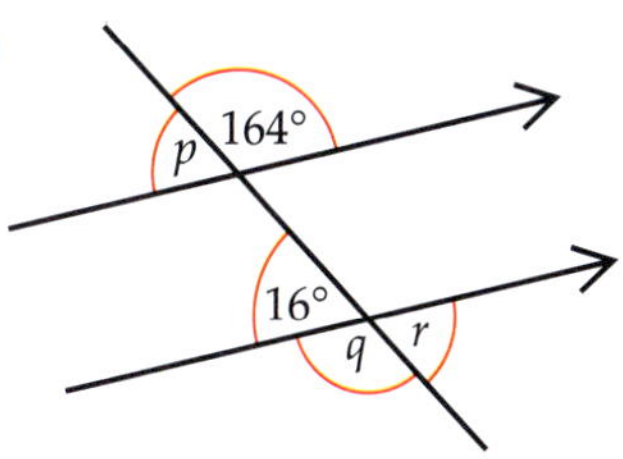

b)

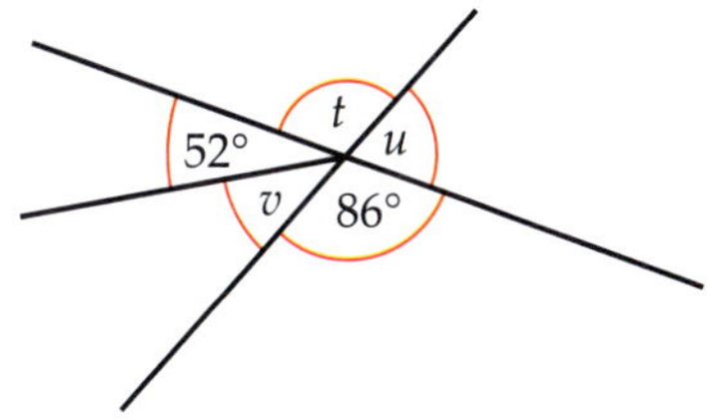

c)

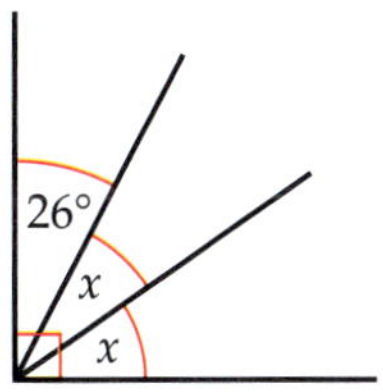

d)

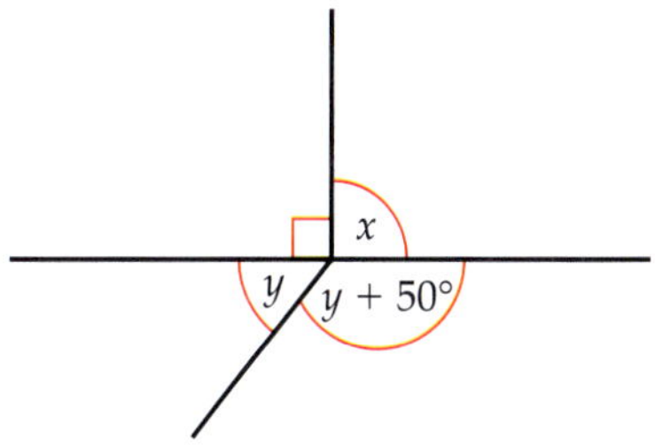

e)

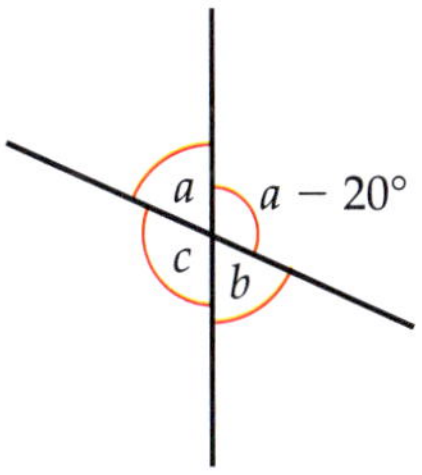

f)

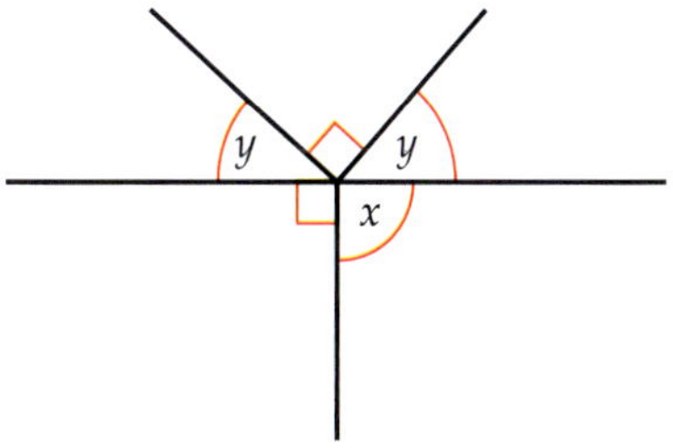

Chapter 12

Solids

This chapter deals with three-dimensional (3-D) or solid objects and how to classify them using their properties. You will learn to sketch 2-D representations of 3-D shapes and how to use nets to construct solids.

Key words
Faces
Edges
Vertex/vertices
2-D 3-D
Cross-section
Prism

Unit 12.1 Describing and classifying shapes

In this unit you will:

- describe solids in terms of their edges, faces and vertices
- use the properties of solids to classify and name them
- sketch 2-D representations of 3-D objects.

Everything around us has a shape. The mathematics of shape and space is called geometry. Shapes which have length, breadth and height are called solids, or three-dimensional (**3-D**) shapes.

Each part of a solid has a special name.

A **face** is the surface of a solid.
Faces can be flat or curved.

An **edge** is the line where two faces meet.
Edges can be straight or curved.

A **vertex** is a point where edges meet.
The plural of vertex is **vertices**.

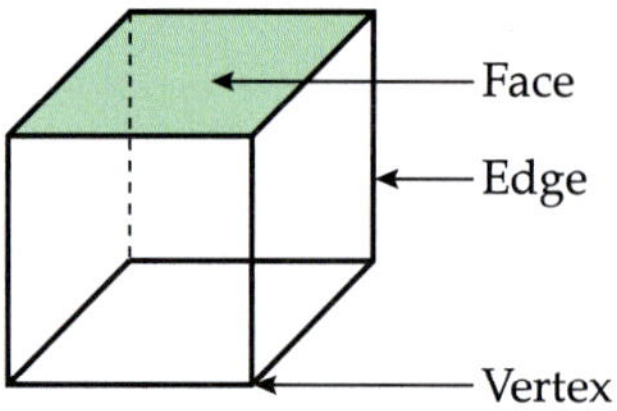

Exercise

1. Find five solids in your environment.
   a) Show a partner the faces, edges and vertices of the objects you find.
   b) Try to find an object with a curved edge. How many vertices does it have? How many faces does it have?

2. Draw a sketch of each object you found.

The basic geometric solids

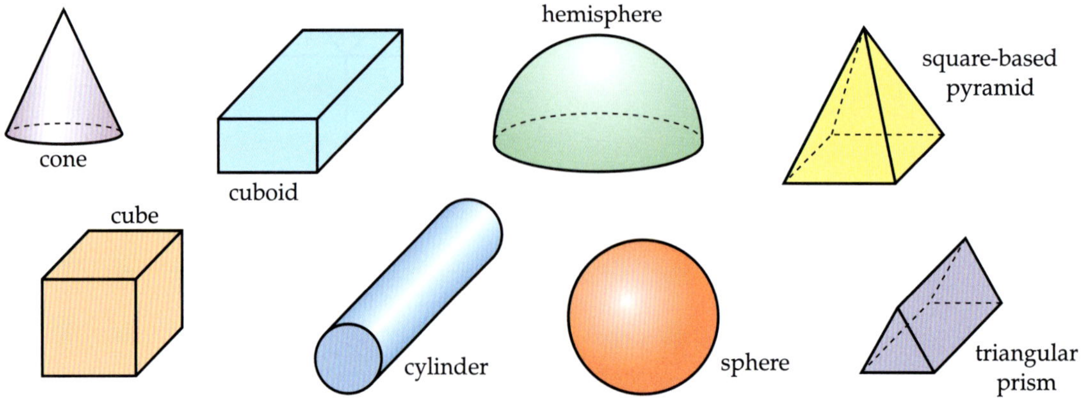

Drawing solids

Pictures and drawings of three-dimensional objects are two-dimensional (**2-D**) representations of the objects. Drawings have length and breadth, but they do not have height. In other words, they are flat on the page.

When you draw 2-D representations of solids you should note that:

- We can use other shapes to represent the solid objects. For example, we use rhombuses to show the faces of a cube.
- Broken lines are used to show hidden edges and vertices.

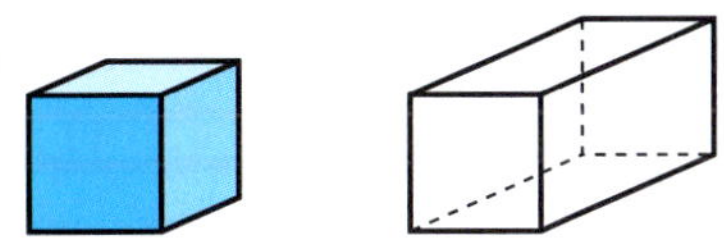

Exercise

1. Collect a number of solid shapes. Use them to find the information you need to complete the table.

Solid	Number of faces	Number of edges	Number of vertices
Cone			
Cuboid			
Cube			
Cylinder			None
Square-based pyramid			
Hemisphere			
Sphere			

Cross-sections of pyramids and prisms

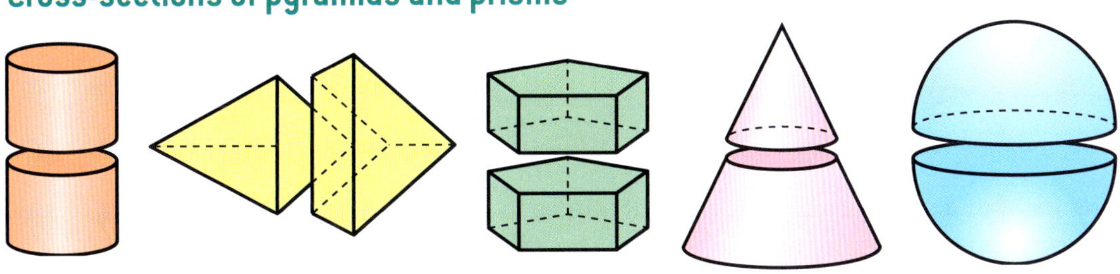

Cutting through an object gives you a flat surface known as a **cross-section**. Some solids have cross-sections that are equal in shape and size. The cross-sections of a pyramid are the same shape, but they differ in size.

Solids which have identical faces which are parallel are called **prisms**. All cross-sections along a prism will produce identical faces.

Exercise

1. Use clay (plasticine) to model at least four different solids. Cut your models to form cross-sections.
   - **a)** Draw the cross-sections of each shape.
   - **b)** Which of your shapes were prisms?

Unit 12.2 Planes and nets

In this unit you will:

- describe the relationship between edges and faces of solids
- recognise the nets of various solids
- draw nets and use them to build solids.

Key words
Planes
Parallel
Perpendicular
Nets
Fold lines

Planes are flat surfaces. They may be horizontal, vertical or slanted. Plane figures are bounded by straight or curved lines.

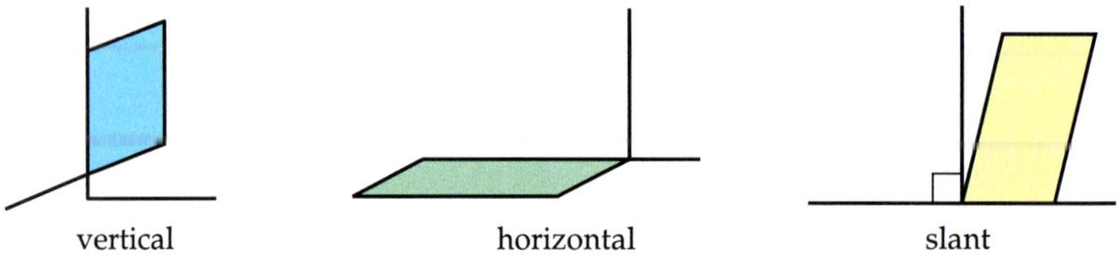

Look around. How many planes can you find? For example:

- The top of a table is a horizontal plane.
- The walls of the room are vertical planes.
- The face of a computer monitor is a slanted plane.

Parallel and perpendicular

You learned earlier that a prism has two identical faces which are **parallel**. Parallel means the same distance apart at all points. The shaded faces of each object are parallel to each other.

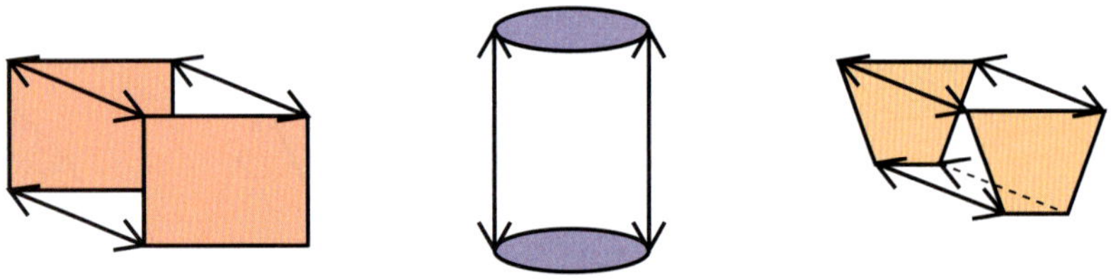

The parts of a solid can also be at right angles to each other. For example, the edges of a cube form right angles at the vertices. The adjacent faces of a cube are also at right angles to each other. Edges and faces at right angles are **perpendicular** to each other.

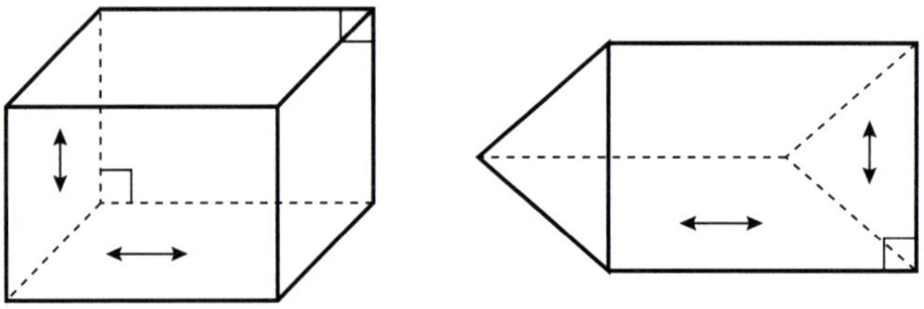

Exercise

1. Find ten examples of parallel and perpendicular surfaces in your environment.

   List where you found each one.

Nets of solids

You can fold some 2-D shapes to make 3-D objects. A two-dimensional shape that is folded to make a three-dimensional shape is called a **net**.

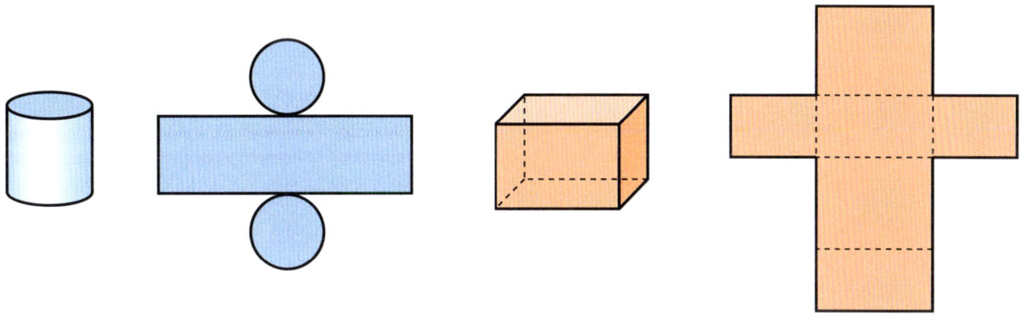

The dotted lines on the nets above are called **fold lines**. The net is folded along these lines to form some of the edges of the solid. If you look at the nets, you should see that each section of the net forms a face of the solid next to it.

Exercise

1. Match each solid to its net. (They are not always the same size.)

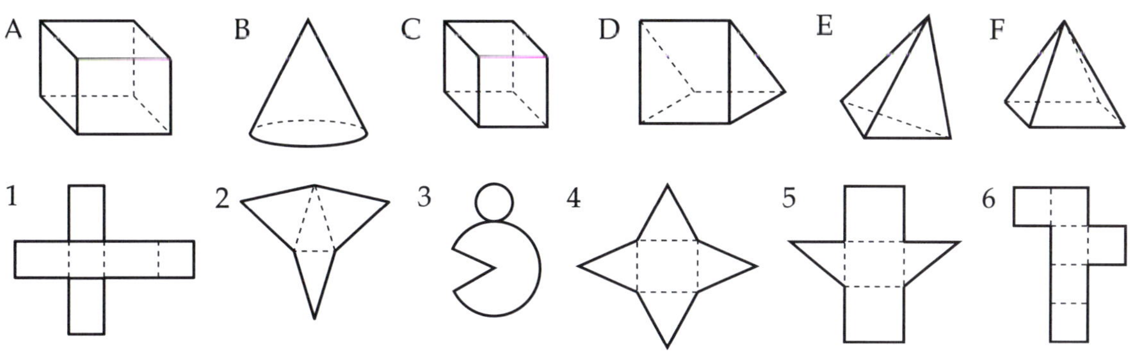

2. Judy built these shapes from nets.
Sketch the nets that you think she used.

3. Joe is a metal worker. He uses nets to make the solids he needs to build things. Draw the nets that you think he used to make this candle-holder.

Looking back

1 Match each solid to a name.

a) **b)** **c)** **d)** **e)** **f)** **g)** **h)**

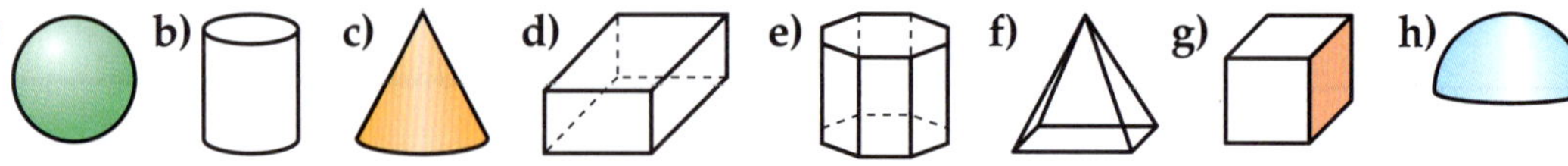

cylinder	cone	cuboid	pyramid	prism	sphere	hemisphere	cube

2

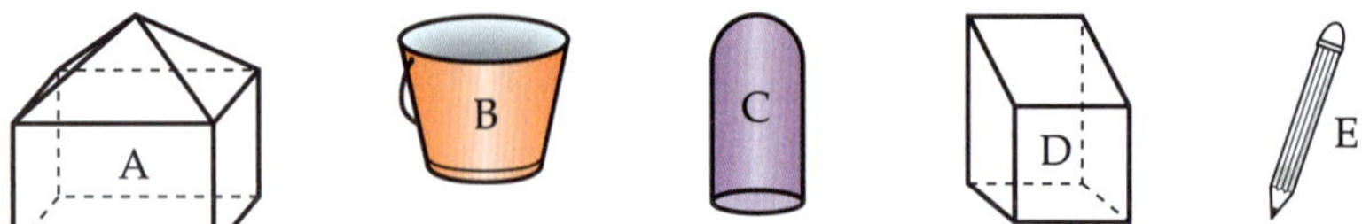

a) Write the letters A to E in your book. Next to each one, write down the number of faces, edges and vertices the shape has.

b) Name the basic solids used to make up each shape.

c) How many curved faces are there in each shape?

3 Sketch the solids you can build from these nets.

a)

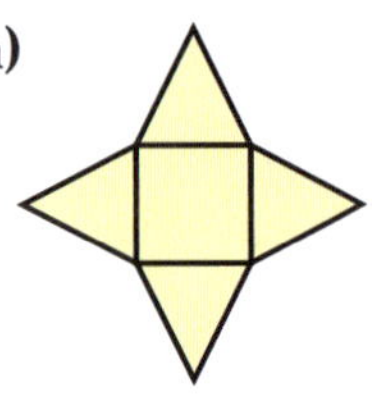

b)

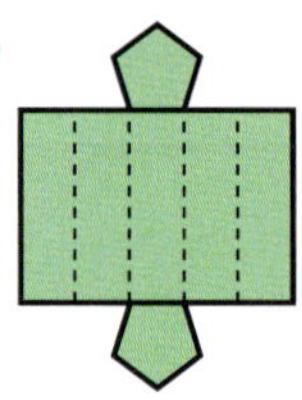

c)

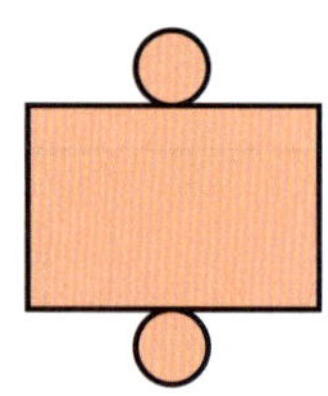

4 The following shapes were made using beads for vertices and straws for edges. Use the diagrams to help you complete the table.

Shape	Number of straws	Number of beads
A 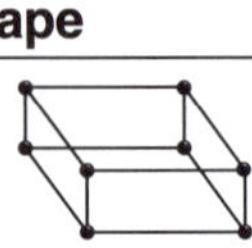		
B 		
C 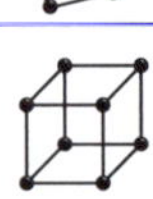		
D 		
E 		

5 Say whether these diagrams show lines or surfaces that are parallel, perpendicular or neither.

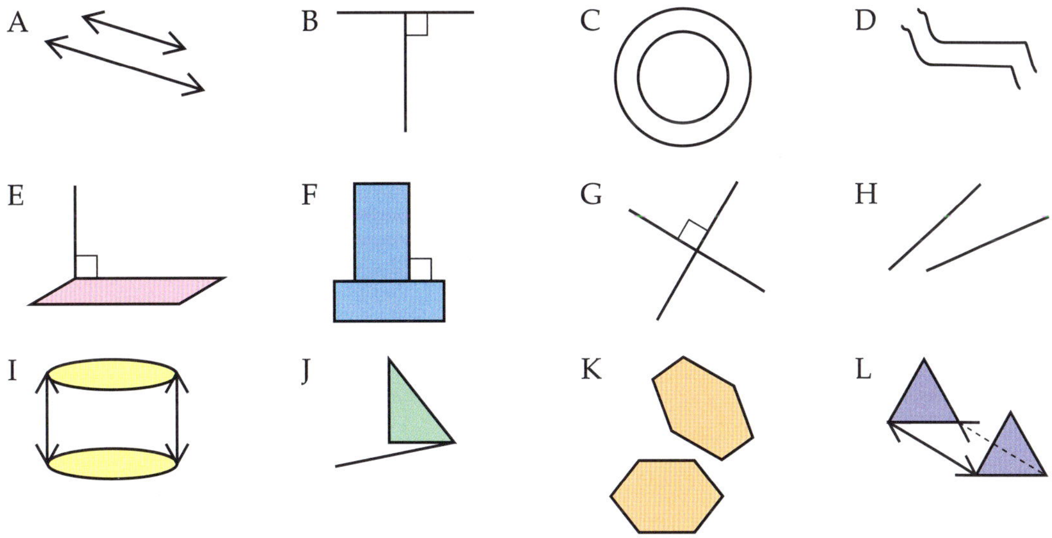

Project

1 Choose three objects in your environment. You can choose buildings, landmarks or everyday objects. However, your objects must be a combination of at least two different solids. Make sketches of the objects you have chosen to show what shapes were used to build them. For example:

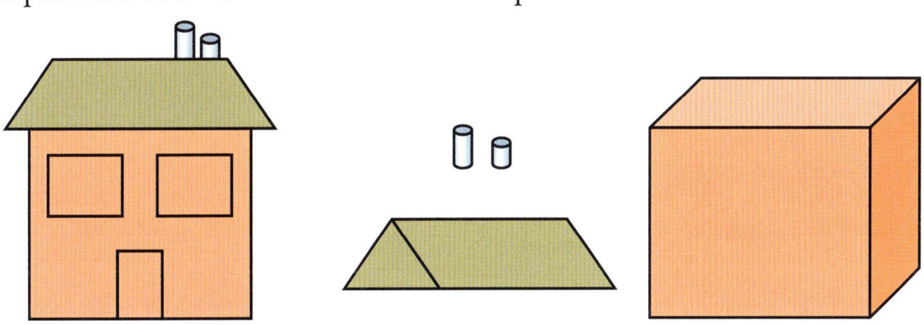

2 Build these solids. Use Bristol paper or thin cardboard to make the nets you need.

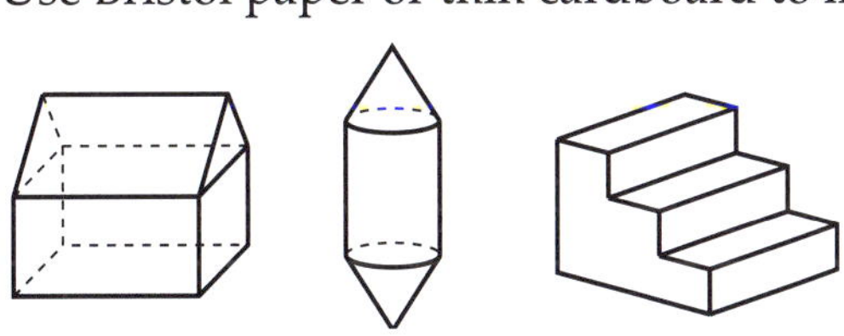

3 Choose one solid you have studied. Write a description of it on a piece of card. Include information about:

a) the number of faces, vertices and edges

b) the shape of the faces

c) whether the surfaces are curved or flat.

When you have done this, place all the cards in a pile and take turns to read the descriptions and name the solids.

Chapter 13

Triangles and quadrilaterals

This chapter deals with different types of triangles and quadrilaterals (plane shapes) that you will come into contact with in mathematics and everyday life. You will examine shapes and learn more about the properties of different shapes. These properties will allow you to use the sides, angles and lines of symmetry to describe and classify shapes and to solve problems. You will also learn to construct lines and angles accurately using a ruler, protractor and compasses.

Key words
Scalene
Isosceles
Equilateral
Acute-angled
Obtuse-angled
Right-angled

Unit 13.1 Triangles

In this unit you will:

- identify and name different kinds of triangles
- use angles and sides to classify triangles.

Triangles can be named according to the number of equal sides they have:

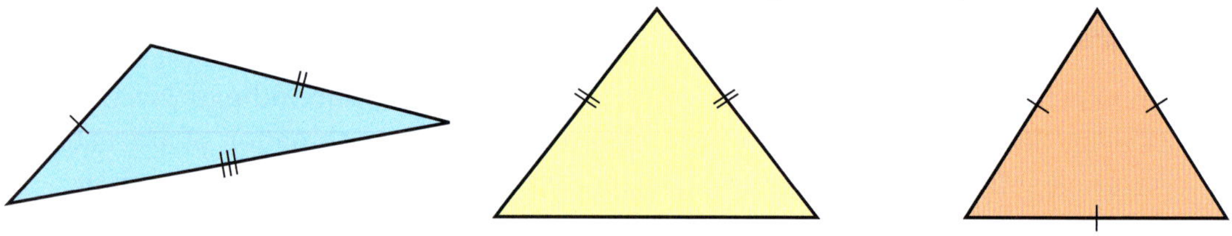

Scalene Isosceles Equilateral

A **scalene** triangle has no equal sides.
An **isosceles** triangle has two equal sides.
An **equilateral** triangle has three equal sides.

Note that sides are marked with dashes. Sides marked with an equal number of dashes are the same length. Sides marked with different numbers of dashes are different lengths.

Triangles can also be classified according to the types of angles they contain:

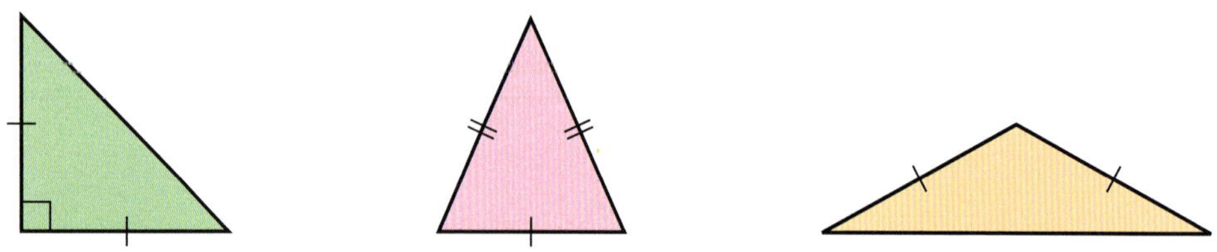

Right-angled Acute-angled Obtuse-angled

An **acute-angled** triangle has three angles that are acute (less than 90°).
An **obtuse-angled** triangle contains one obtuse angle (greater than 90° but less than 180°).

A **right-angled** triangle contains one right angle (90°).

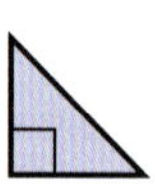

Note that we use a small square to mark right angles.

Worked examples

1. Classify these triangles using their sides:

a. 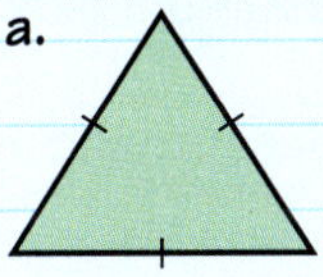

b. 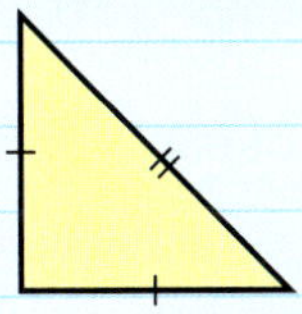

c. 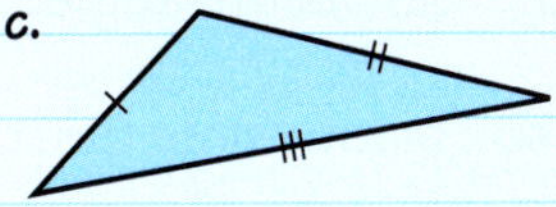

2. Use angle sizes to classify these triangles:

a. 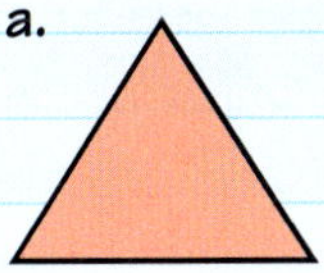

b. 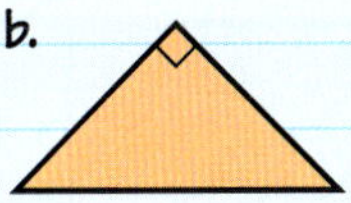

c. 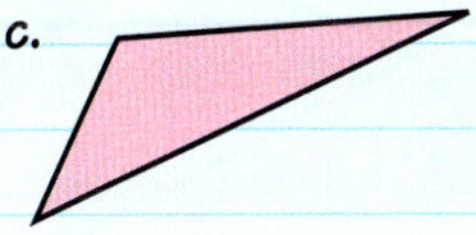

3. Use sides and angles to name these triangles:

a. 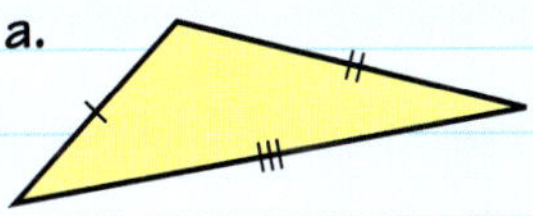

b. 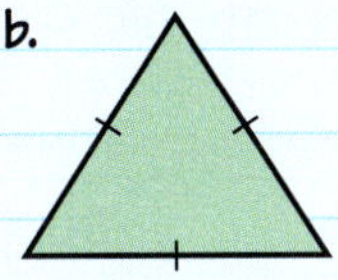

c. 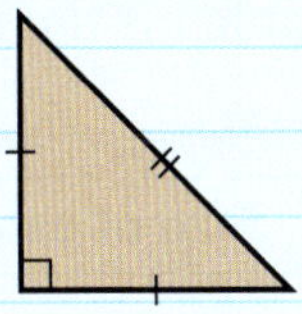

1. a. Three sides are equal so the triangle is equilateral.
   b. Two sides are equal so the triangle is isosceles.
   c. No sides are equal so the triangle is scalene.
2. a. All angles are acute so the triangle is acute-angled.
   b. There is a right angle, so the triangle is right-angled.
   c. One angle is obtuse, so the triangle is obtuse-angled.
3. a. No sides are equal and there is an obtuse angle, so the triangle is an obtuse-angled scalene.
   b. All sides are equal and all angles are acute, so the triangle is an acute-angled equilateral.
   c. Two sides are equal and there is a right angle, so the triangle is a right-angled isosceles.

Exercise

1 Use a ruler to measure the side lengths of the following triangles and use your measurements to name each triangle:

a) 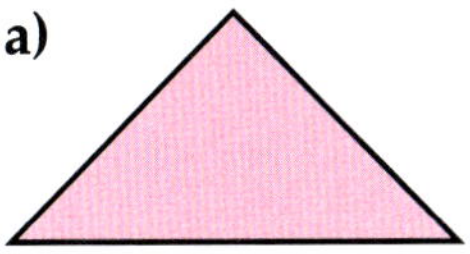

b) 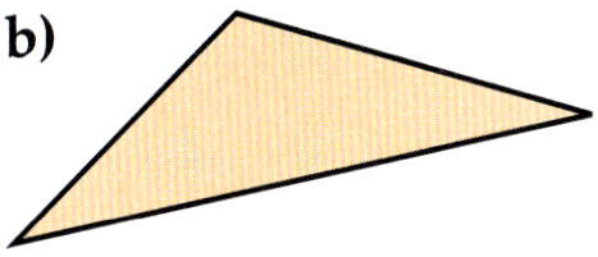

c) 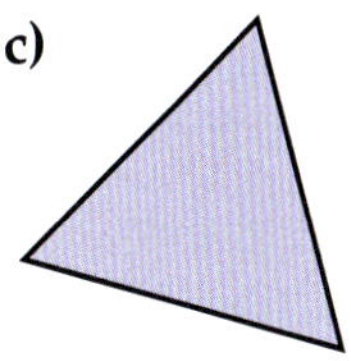

d) 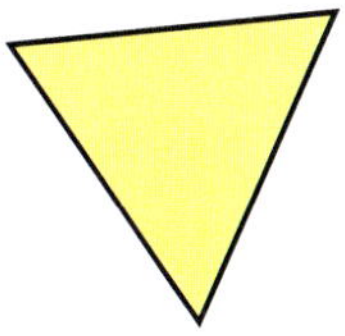

e) 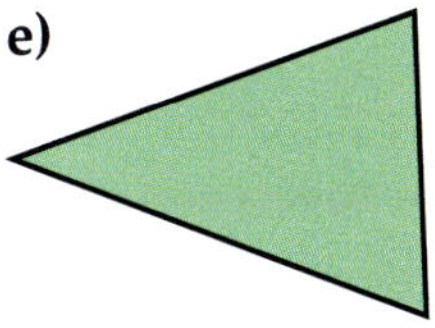

f) 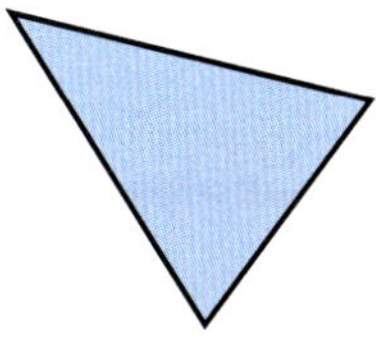

g) 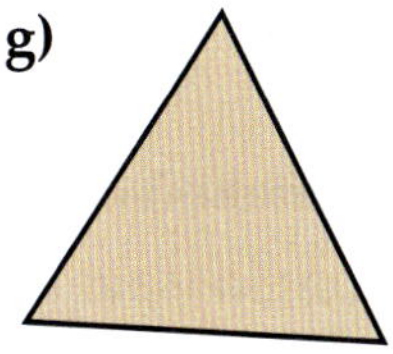

h) 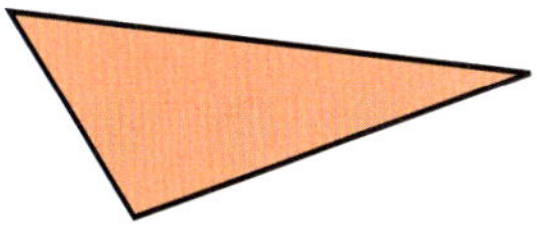

2 Sketch and label a triangle that is:

a) right-angled **b)** right-angled isosceles **c)** scalene obtuse-angled

d) scalene acute-angled **e)** acute-angled isosceles **f)** obtuse-angled isosceles

3 For each triangle in the box below, state the name of each triangle. Measure the angles if necessary.

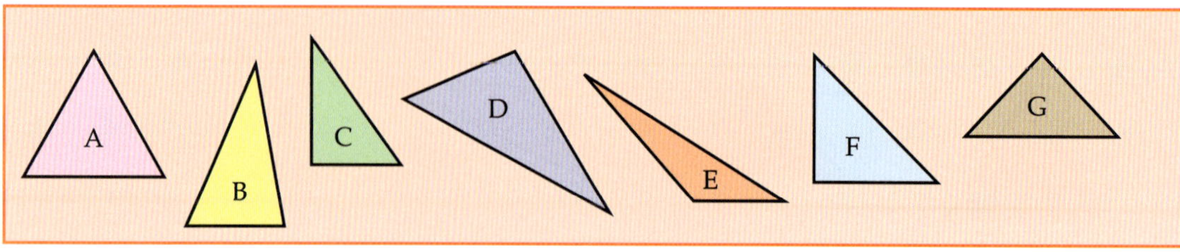

4 Copy and complete this table for the named triangles, suggesting values for each angle in the triangle.

Name of triangle	Size of each angle	Sum of the angles
Right-angled		
Right-angled isosceles	90°, …, …	
Scalene obtuse		
Scalene acute		
Acute isosceles		
Obtuse isosceles		
Equilateral	60°, 60°, 60°	

Investigation

1 How many different (in shape and size) triangles can you draw on a grid of nine dots like the one below? Each vertex of your triangles must lie on a dot.

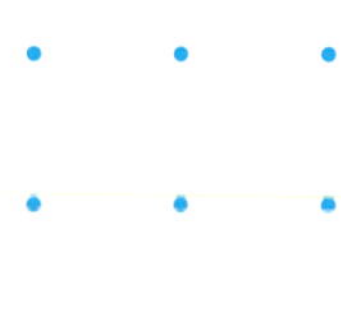

2 Make the following patterns with spent matches or toothpicks. Follow the instructions for each one:

a)

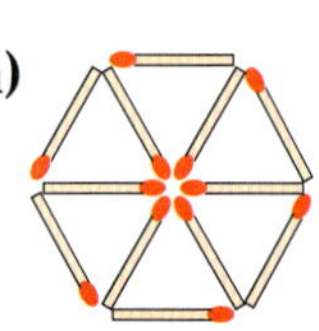

b)

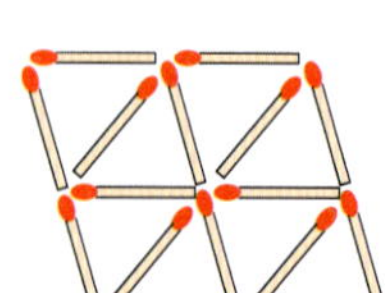

a) Move 4 matches to make 3 equilateral triangles.

b) Take away 4 matches to leave 4 congruent equilateral triangles.

Key words
Interior
Exterior

Unit 13.2 Calculating angles

In this unit you will:

- know and use the fact that the sum of the angles in a triangle is 180°
- know and use the fact that the sum of the angles in a quadrilateral is 360°
- understand that the exterior angle of a triangle is equal to the sum of the two interior opposite angles.

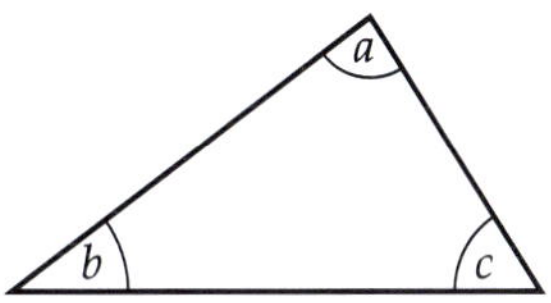

$a + b + c = 180°$ (angles on a straight line)

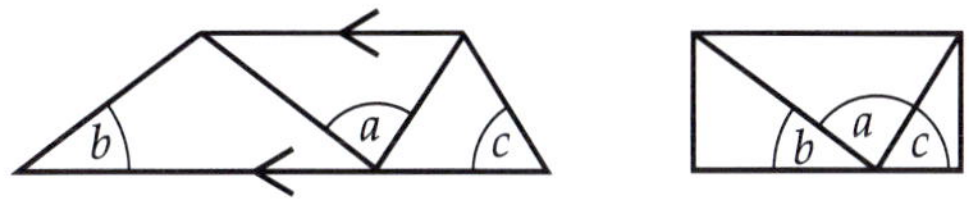

Folding the corners of a triangle shows you that the angles make a straight line. This means that the sum of the angles in any triangle is 180°.

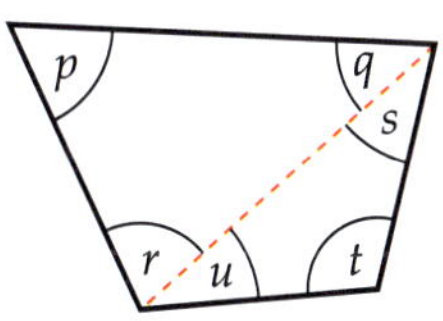

$p + q + r + s + t + u = 360°$

All quadrilaterals can be split into two triangles. The sum of the angles in any quadrilateral is $2 \times 180° = 360°$.

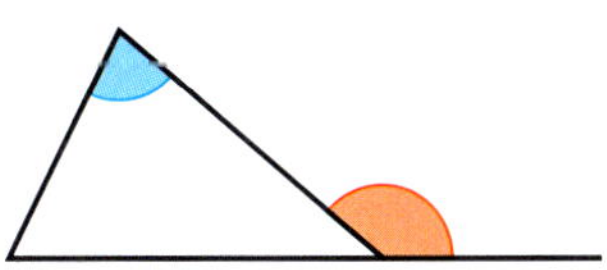

An **interior** angle is *inside* the shape.
An **exterior** angle is *outside* the shape. It is made by extending one of the lines.

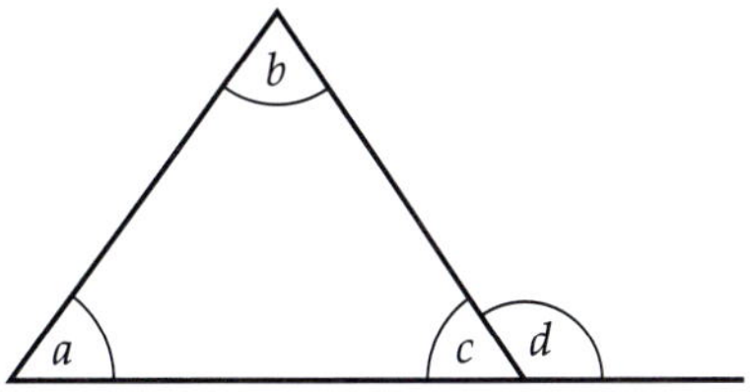

$c + d = 180°$ (angles on a straight line)
$a + b + c = 180°$ (angles in a triangle)
so $a + b = d$

The exterior angle of a triangle is equal to the sum of the two interior opposite angles.

Worked example

Calculate the size of the lettered angles, giving reasons for your answers.

a.

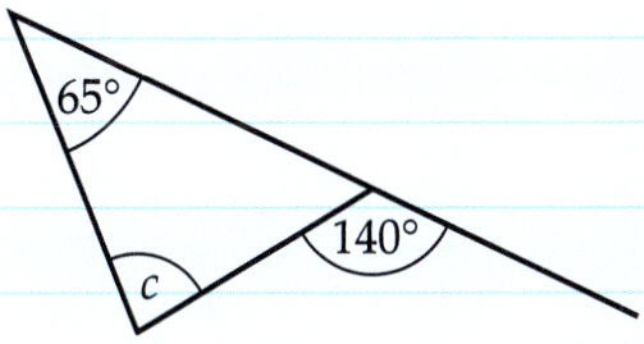

b.

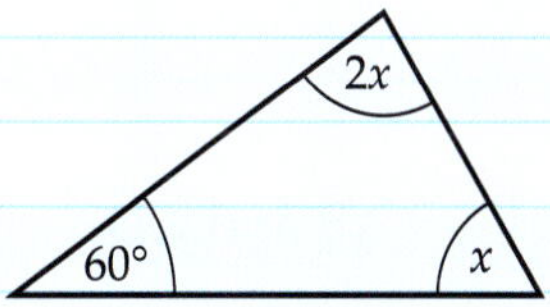

a. $c + 65° = 140°$
$c = 140° - 65°$
$= 75°$

The exterior angle of a triangle is equal to the sum of the two interior opposite angles.

b. $2x + x + 60° = 180°$
$3x + 60° = 180°$
$3x = 120°$
$x = 40°$ and $2x = 80°$

The sum of angles in a triangle is 180°.

Exercise

1 Calculate the size of the lettered angles, giving reasons for your answers.

a)
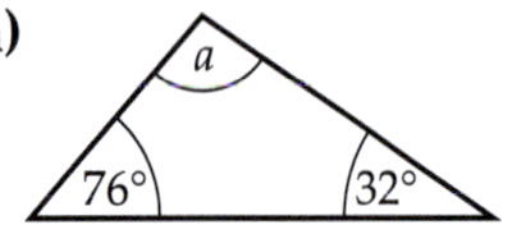

b)
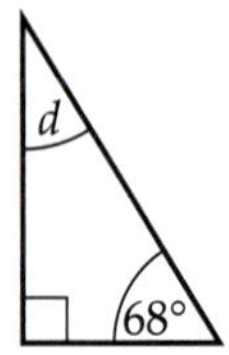

c)
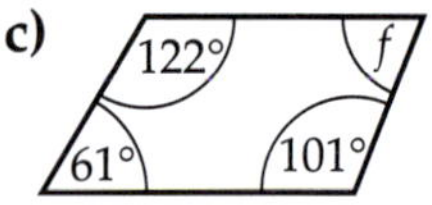

d)
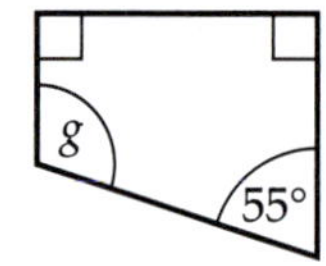

e)
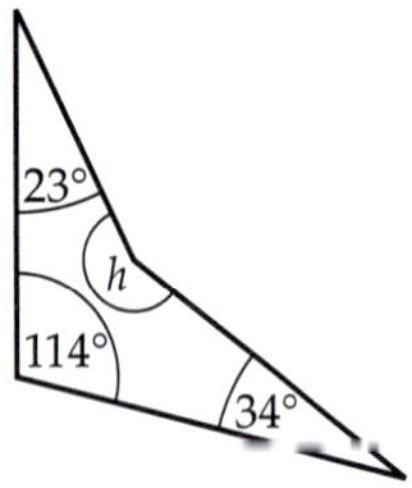

f)
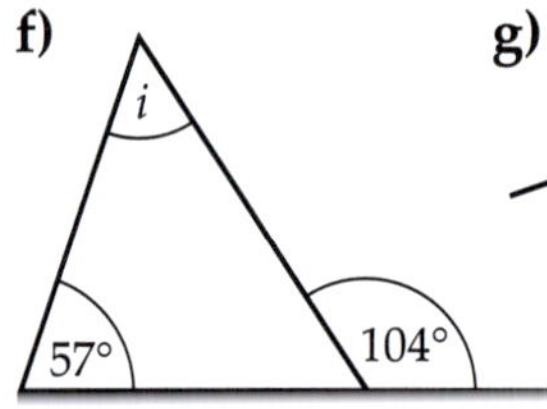

g)
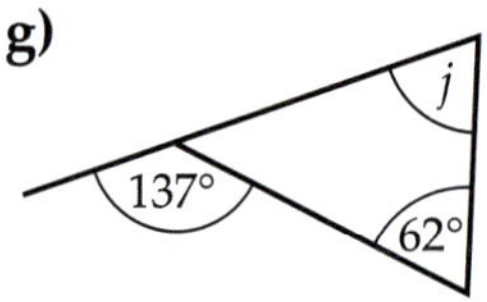

h)
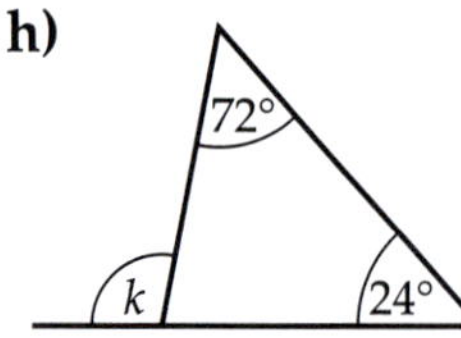

2 Calculate the size of the lettered angles, giving reasons for your answers.

a)
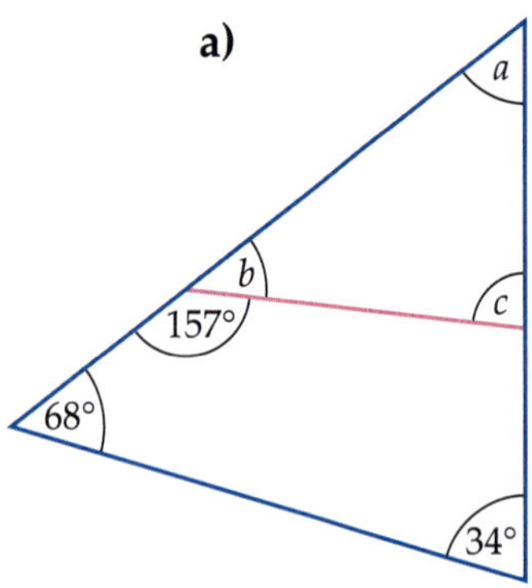

Look at the blue triangle first and find *a*.

b)
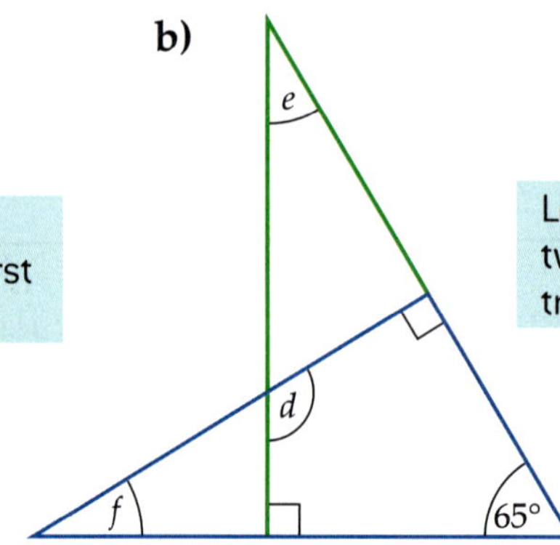

Look for the two large triangles.

3 Work out the value of x in each of the following:

a)
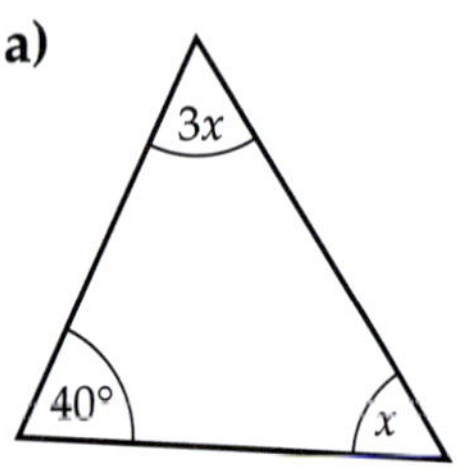

b)
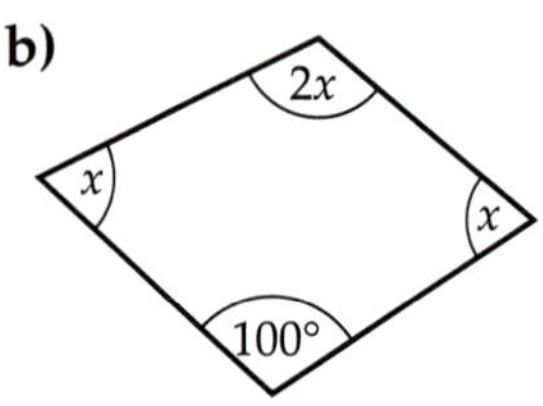

c)
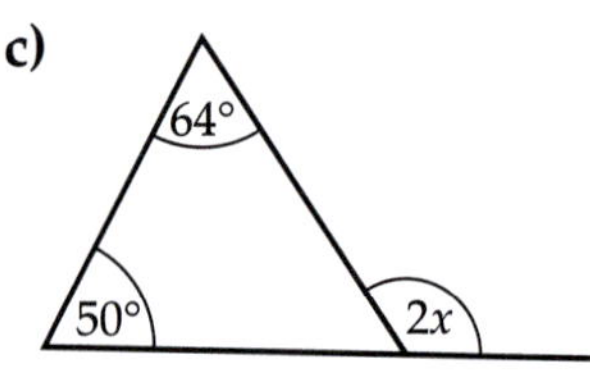

Investigation

Draw two straight lines AB and CD.
Mark a point O between them.
Draw a line through O that crosses the two lines.
Mark the two angles, underneath the line, as x and y.
Measure the size of x and y with a protractor.
Draw some more lines through O. Each time measure the angles underneath the line and record your results in a table.
What do you notice? Can you explain why this happens?

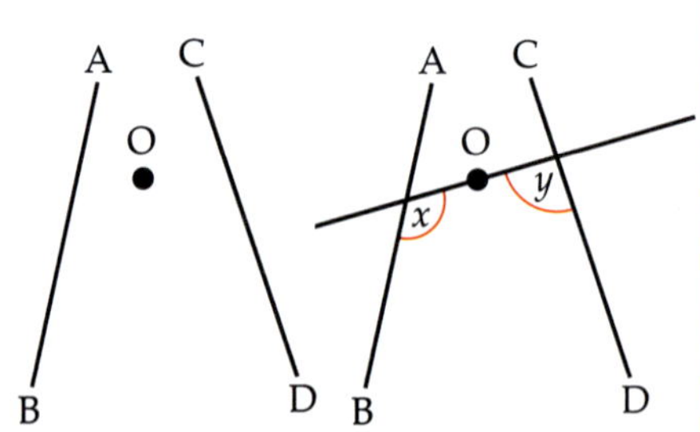

Extend AB and CD so they meet.

Unit 13.3 Quadrilaterals

In this unit you will:

- identify and name different kinds of quadrilaterals
- use angles and sides to classify quadrilaterals.

A **quadrilateral** is a 4-sided shape. We name quadrilaterals by looking at the properties of their angles and sides. These are examples of quadrilaterals with the properties that give them their names.

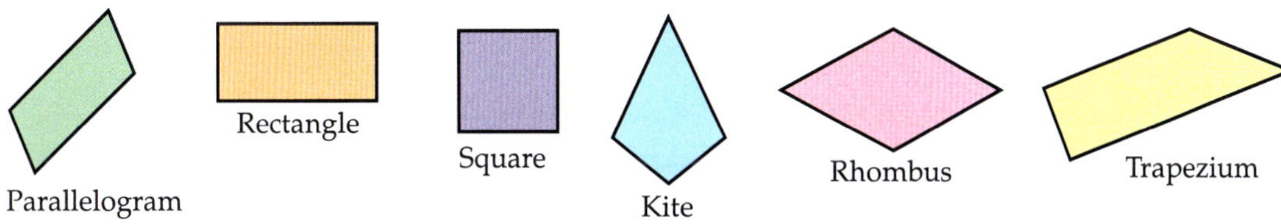

Quadrilateral
Any 4-sided shape.

Parallelogram
Opposite angles are equal, opposite sides are equal and parallel.

Rhombus
Opposite angles equal, four equal sides, opposite sides parallel.
A rhombus is a parallelogram with all sides equal.

Kite
Two pairs of adjacent equal sides, one pair of opposite angles are equal.

Rectangle
All angles 90°, opposite sides are equal and parallel.
A rectangle is a parallelogram with 90° angles.

Square
All angles 90°, all sides equal, opposite sides parallel.
A square is a rectangle with all sides equal.

Trapezium
One pair of opposite sides parallel.

Any line from vertex to vertex across a quadrilateral is called a **diagonal**.

A particular quadrilateral may have more than one correct name. For example, a square may also correctly be called a rectangle. We use the name square to be more precise and because all squares are rectangles, but not all rectangles are squares.

Exercise

1. Give the most precise name for each of the shapes below:

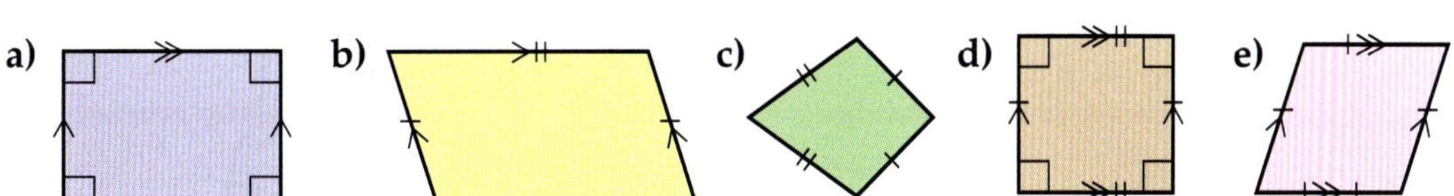

2 Copy the tables. Use your knowledge of quadrilaterals to complete them.

Quadrilateral	Opposite sides are always parallel and equal (Yes/No)	Opposite angles are always equal (Yes/No)	Angles are always right angles (Yes/No)
Parallelogram			No
Rectangle			
Square	Yes		Yes
Rhombus			
Kite	No		
Trapezium	No		

3 What type of quadrilateral is each of these?

a) two pairs of parallel sides; four equal sides; four right angles

b) one pair of parallel sides; no equal sides; no right angles

c) two pairs of parallel sides; four equal sides; no right angles

d) no parallel sides; two equal sides; one right angle

4 These definitions are incomplete, so they could apply to more than one quadrilateral. List all the quadrilaterals to which each description could apply.

a) It has four equal sides.

b) It has two pairs of equal sides.

c) It has two pairs of parallel sides.

d) It has one pair of parallel sides and one pair of equal sides.

5 Draw two different examples of each of the following quadrilaterals:

a) a kite
b) a parallelogram
c) a rhombus
d) a rectangle
e) a square
f) a trapezium

6 Say whether the following statements are true or false:

a) The opposite angles of a parallelogram are equal.

b) All four angles in a parallelogram are always equal.

c) All four angles in a rhombus are equal.

d) The opposite angles in a rhombus are equal.

e) One pair of opposite angles in a kite will always be equal.

7 Each of these shapes is half of another. What could the whole shape be? Draw sketches and record your working.

a)

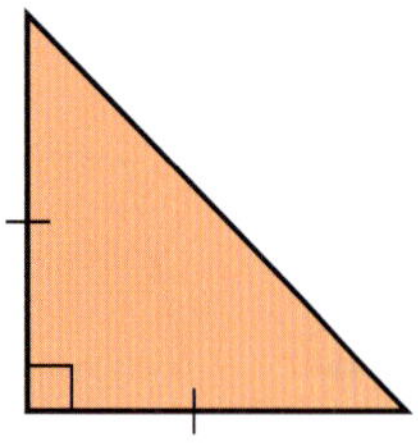

b)

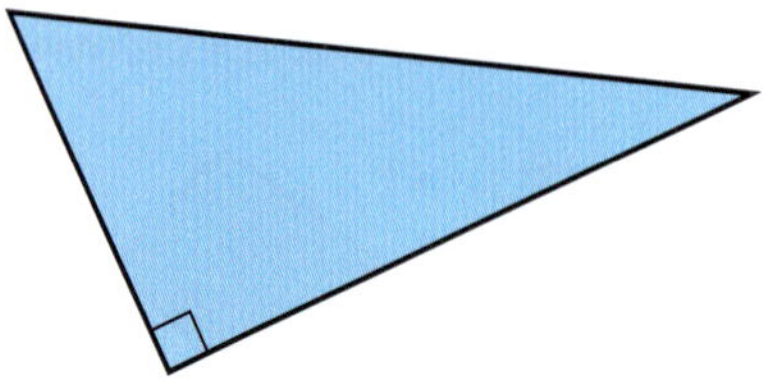

c)

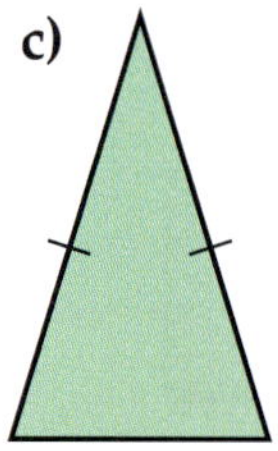

8 The diagrams below are rough sketches, but they are not very accurate.
Given that the markings on the lines and angles are correct, draw more accurate versions of each shape.

a)

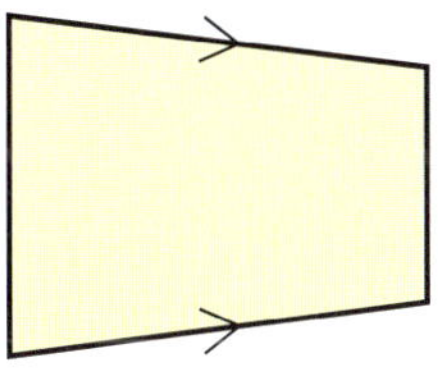

b)

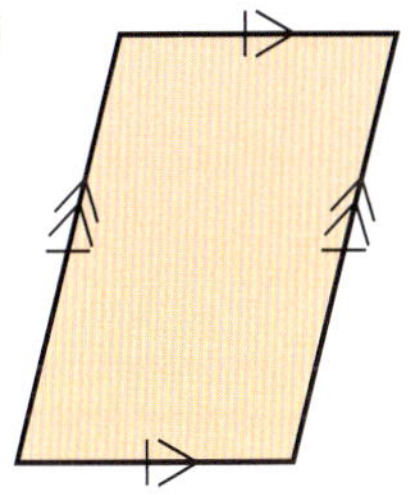

c)

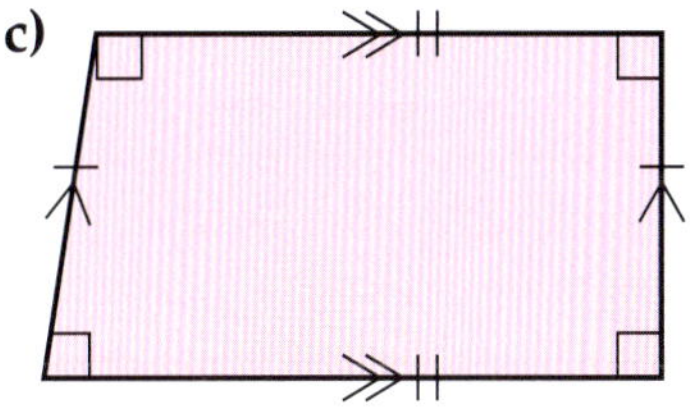

d)

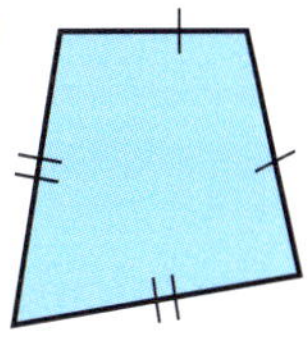

e)

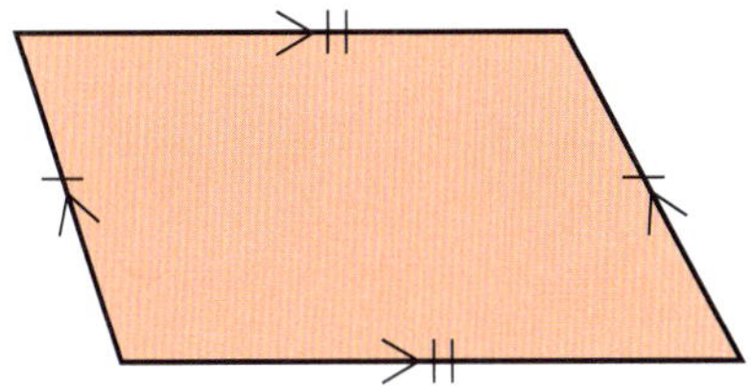

f)

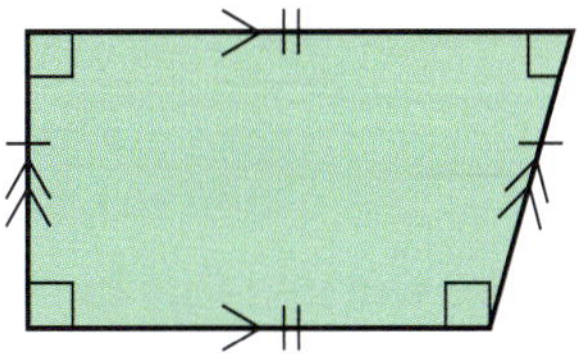

Investigation

1 How many different sized squares can you draw on the dot grid? Each corner of the square must be on a dot.

2 Develop arguments with friends about the quadrilaterals. One example is given.

Question: Is a square a trapezium and a trapezium a square?

Response: A square is a trapezium because if opposite sides are parallel then at least one set of opposite sides are parallel. It is a special trapezium.

However, not all trapeziums have two sets of parallel sides, all sides are not always equal and all the angles are not always equal to 90°, so a trapezium is not always a square.

Unit 13.4 Solving geometrical problems

Key words
Vertically opposite

In this unit you will:

- solve geometrical problems using side and angle properties of special triangles and quadrilaterals.

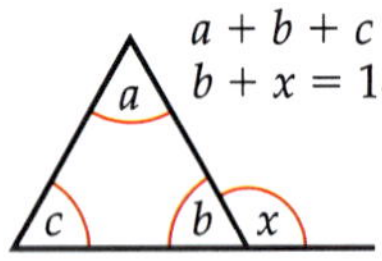

Angles in a triangle add up to 180°.
Angles on a straight line add up to 180°.

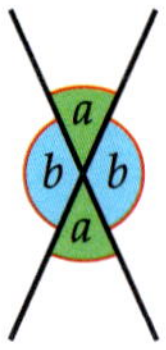

Vertically opposite angles are equal.

We can use this knowledge to find the size of missing angles. Since we are calculating the sizes of the angles by following a chain of reasoning and not measuring them, our diagrams do not have to be accurate.

Worked examples

1. Calculate the size of angles a, b and c, giving reasons for your answers.

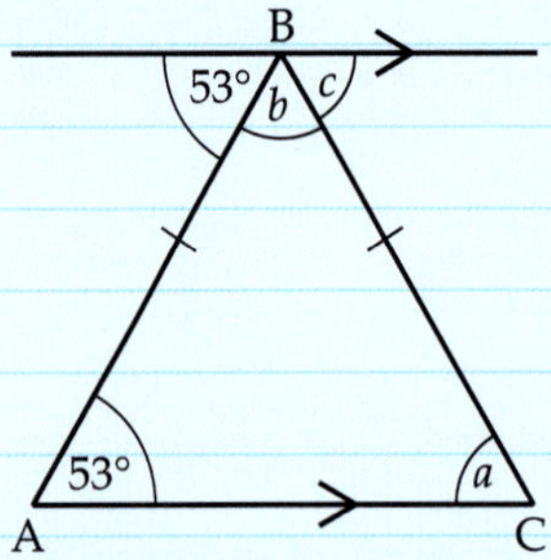

We can see from the diagram that ABC is an isosceles triangle.

Angle a is 53° because triangle ABC is isosceles and the base angles of an isosceles triangle are equal.

Angle b is 74° because angles in a triangle add up to 180° and $180° - 53° - 53° = 74°$.

Angle c is 53° because angles on a straight line add up to 180° and $180° - 53° - 74° = 53°$.

2. Calculate the size of the missing angles giving reasons for your answers.

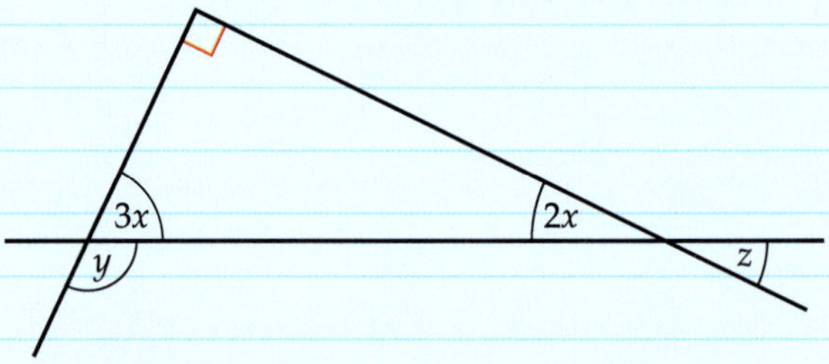

$90° + 3x + 2x = 180°$ because angles in a triangle add up to 180°.

$90° + 5x = 180°$

$5x = 90°$

$x = 18°$ so $3x = 54°$ and $2x = 36°$

$y + 54° = 180°$ because angles on a straight line add up to 180°

so $y = 126°$

$z = 36°$ because it is vertically opposite to the angle 2x.

Exercise

1 Calculate the size of the lettered angles, giving reasons for your answers.

a)

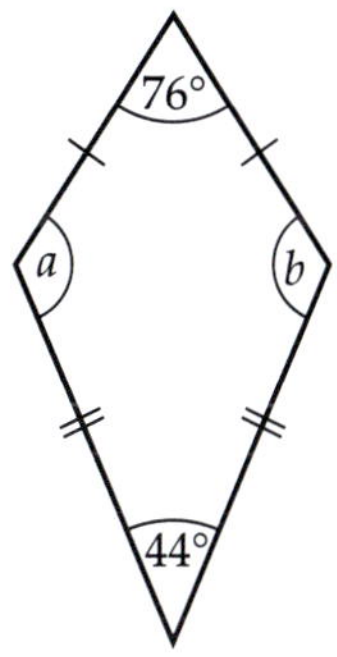

b)

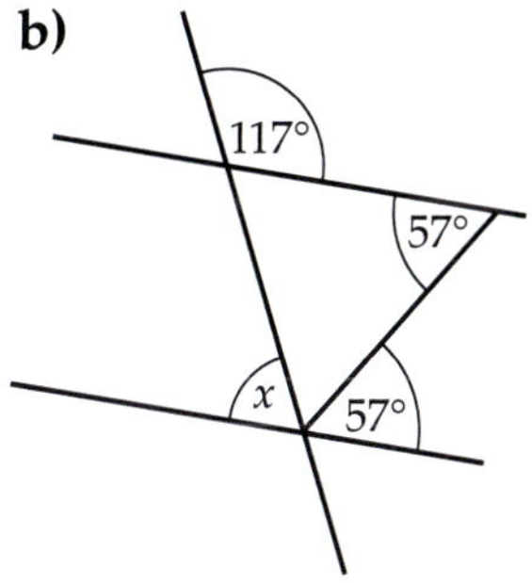

c)

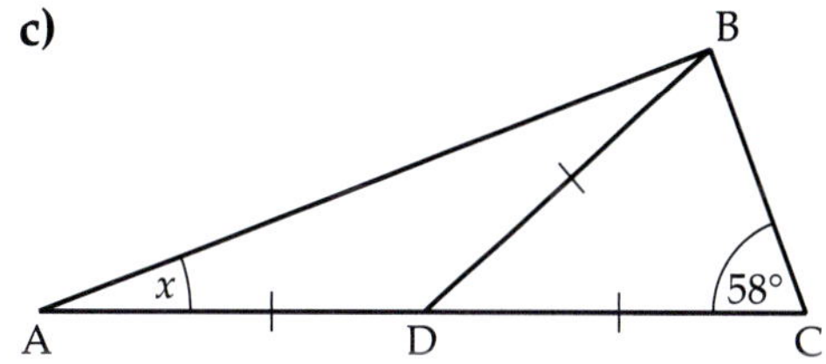

2 Calculate the size of the lettered angles, giving reasons for your answers.

a)

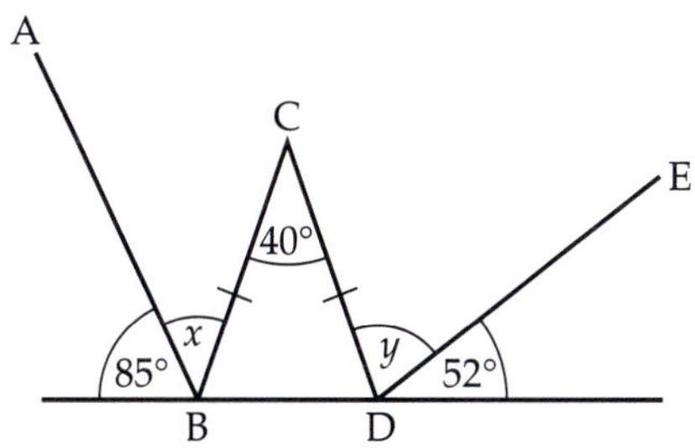

b)

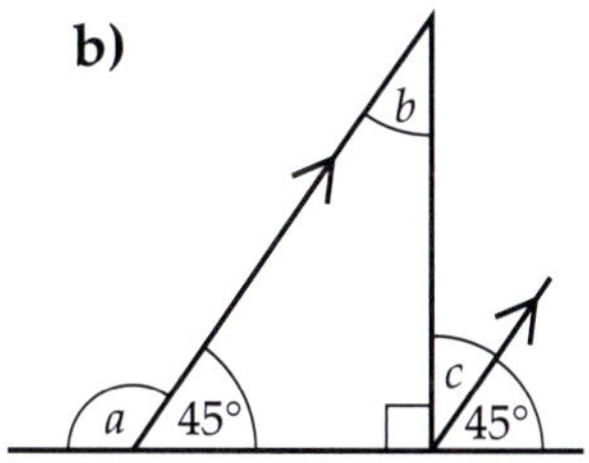

c)

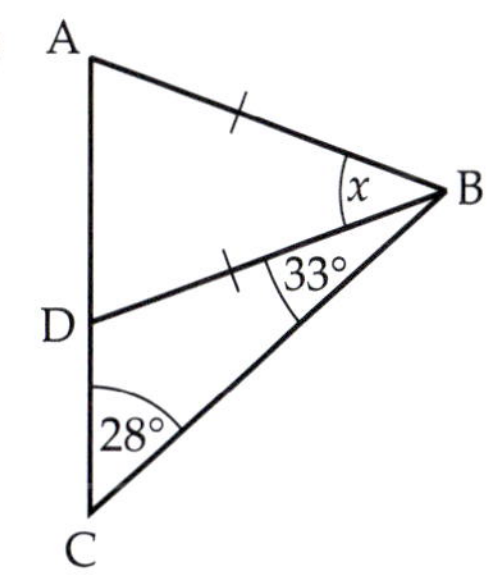

3 Calculate the size of each angle, giving reasons for your answers.

a)

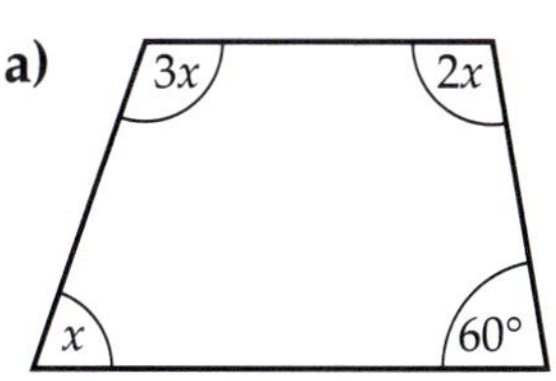

b)

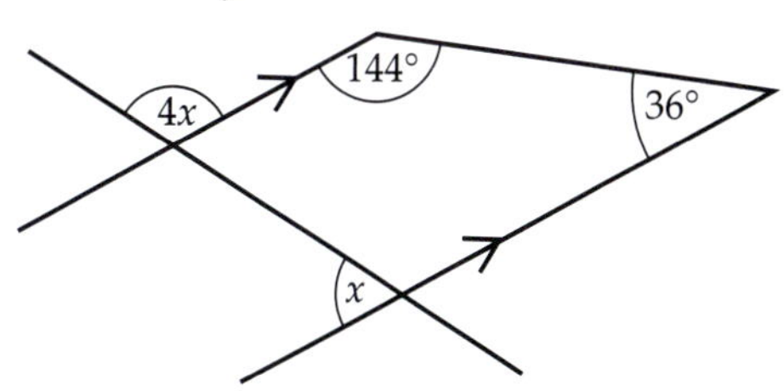

Investigation

The angle at the vertex of a regular pentagon is 108°.
Two diagonals are drawn to the same vertex to make three triangles.

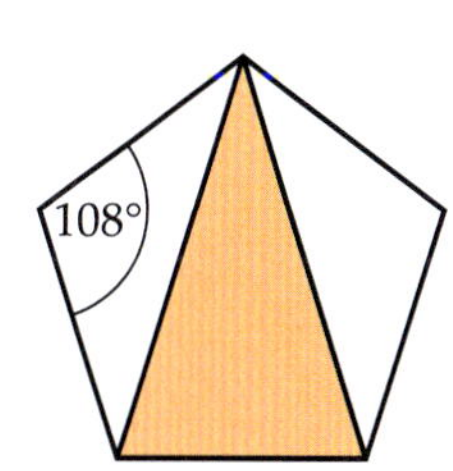

a) Calculate the sizes of the angles in each triangle.

b) The middle triangle and one of the other triangles are placed together like this:

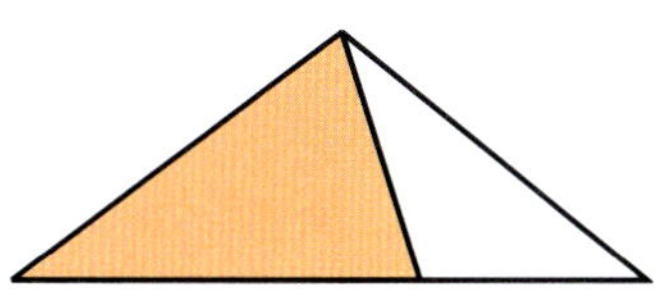

Explain why the triangles fit together to make a new triangle. What are its angles?

c) Investigate the other shapes you can make by putting the three triangles together.

Unit 13.5 Construction

In this unit you will:

- construct the bisector of an angle, using a ruler and compasses
- construct the mid-point and perpendicular bisector of a line segment, using a ruler and compasses.

Key words
Bisector
Compasses
Equidistant
Perpendicular bisector
Mid-point

The **bisector** of an angle is a line that divides the angle into two equal parts. You can construct the bisector of an angle using **compasses**.

In this diagram, BD is the bisector of the angle ABC. Every point on the line BD is **equidistant** from the lines BA and BC.

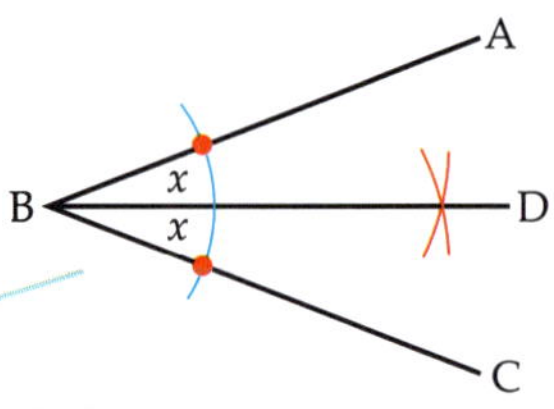

The **perpendicular bisector** of a line segment divides the line segment into two equal parts at right angles.

In this diagram, BD is the perpendicular bisector of AC. It crossses AC at the **mid-point** (M) of the line. Every point on the line BD is equidistant from both A and C. If you join ABCD, a rhombus is formed.

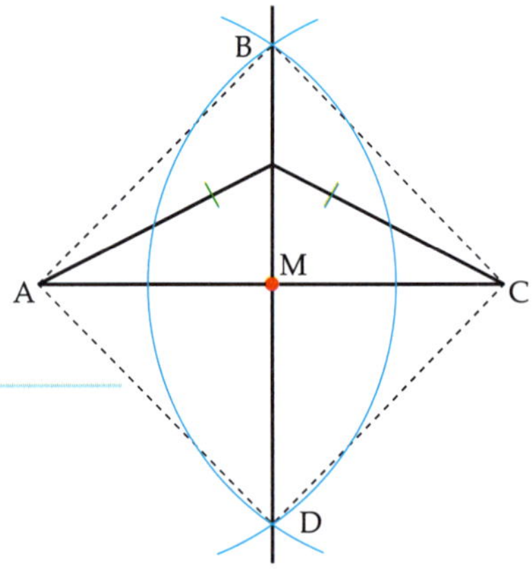

Worked examples

1. ∠DEF is 100°. Construct the bisector of ∠DEF.

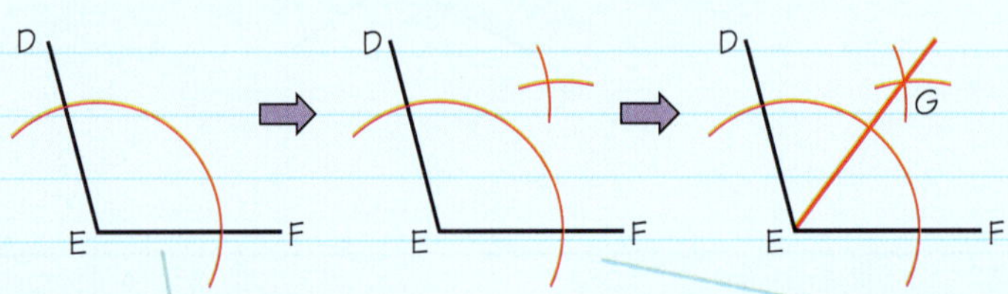

An arc is a part of a circle. You can draw arcs with compasses.

Open the compasses and put the point on E. Draw an arc that intersects with ED and EF.

Do not change the opening of the compasses. Put the point first on the intersection of the arc with ED and then with EF. Draw new arcs to intersect at G. Join EG.

2. The line segment PQ is 2.5 cm long. Construct the perpendicular bisector of PQ.

Draw a line PQ of 2.5 cm. Open up the compasses to over half the length of PQ.

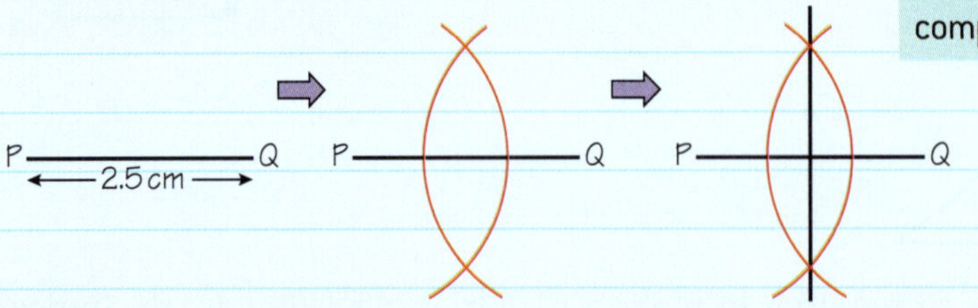

Place the point at P and draw an arc. Keep the opening of the compasses the same and repeat at Q. Join the points where the arcs intersect.

Exercise

1. Draw any acute angle.
Construct the angle bisector using only a ruler and compasses.

2. Draw a line 10 cm in length.
Construct the perpendicular bisector of the line using only a ruler and compasses.
Mark the mid-point (M) of the line.

3. Draw any obtuse angle.
Construct the angle bisector using only a ruler and compasses.
Check it is correct by measuring the two angles and making sure they are the same.

4. A new water channel is to be built at the zoo.
It is to be placed between two animal houses so that the houses are equidistant from the channel.
Copy this diagram, and construct a red line to show the position of the water channnel.

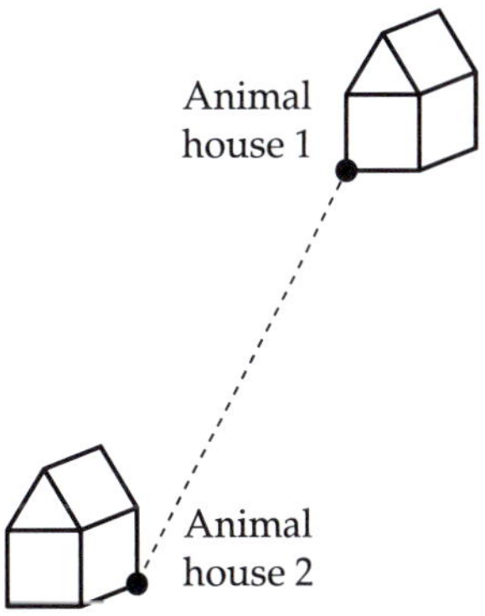

5. A new fence is to be built in the park.
The fence is to be placed between two large trees so that the trees are equidistant from the fence.
The head gardener decides to paint in a line to mark the position of the fence.
Copy this diagram and construct a blue line to show the position of the fence.

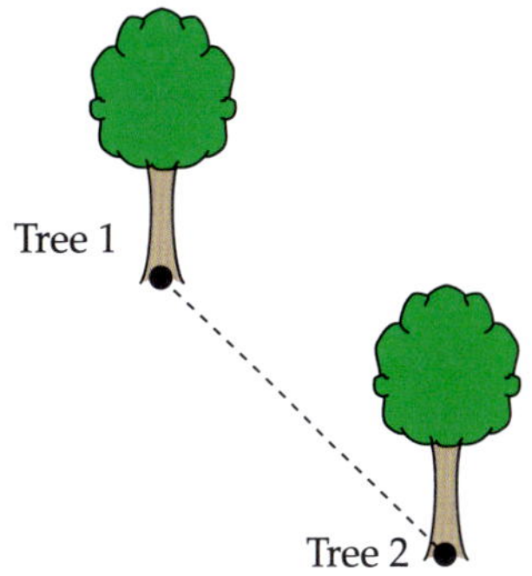

Investigation

a) Draw any triangle with sides longer than 4 cm but shorter than 8 cm. Construct the angle bisector for each of the three angles. Make sure the angle bisectors are long enough to cross each other. What do you notice?

b) Draw another triangle approximately the same size. Construct the perpendicular bisector for each of the three sides. Make sure the perpendicular bisectors are long enough to cross each other. What do you notice?

Looking back

1 State the side and angle name for each of the following triangles. Use a ruler and protractor to make measurements if necessary.

a) 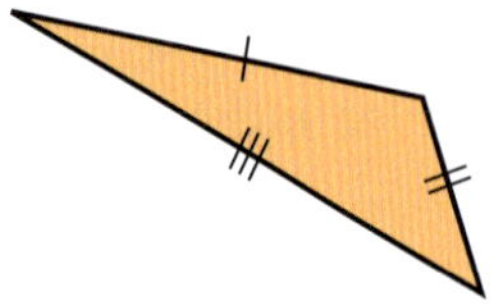

b) 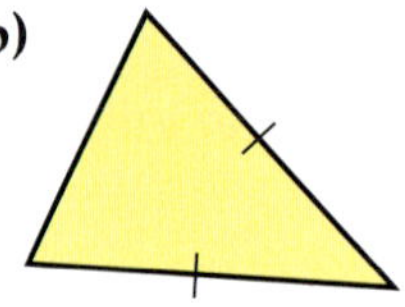

c) 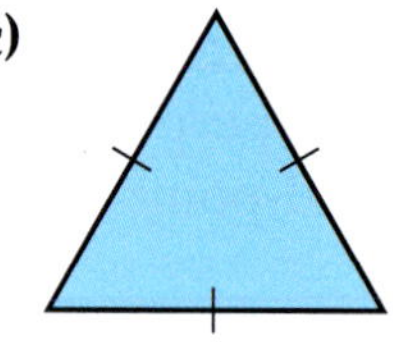

2 Find the value of angle x in each case.

a) 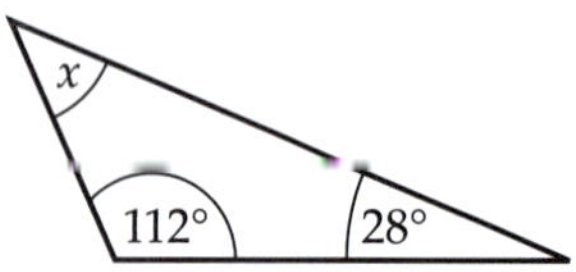

b) 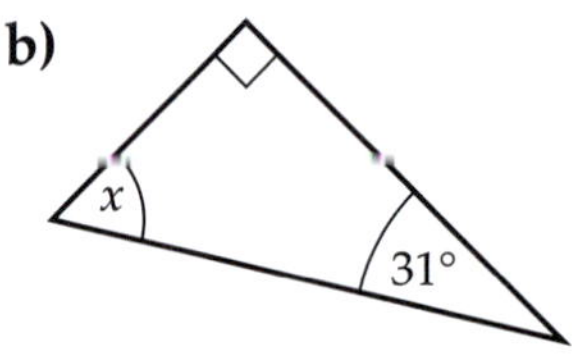

c) 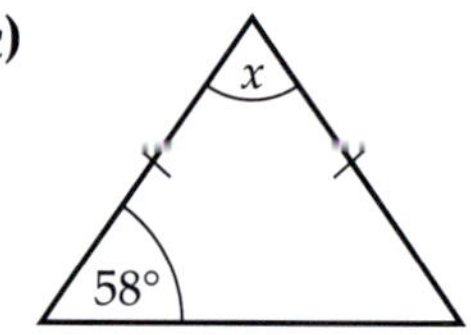

d) 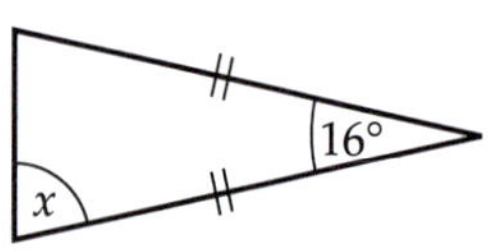

e) 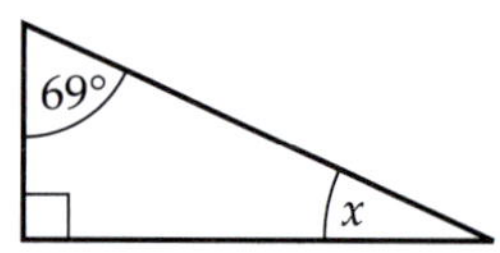

f) 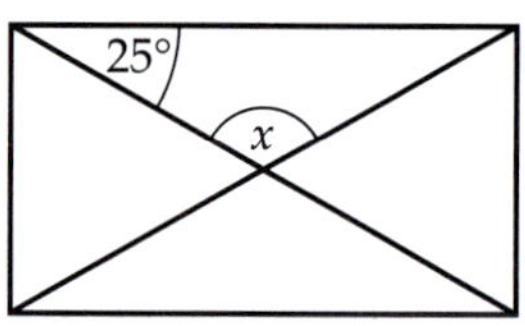

3 Name the following quadrilaterals:

a) 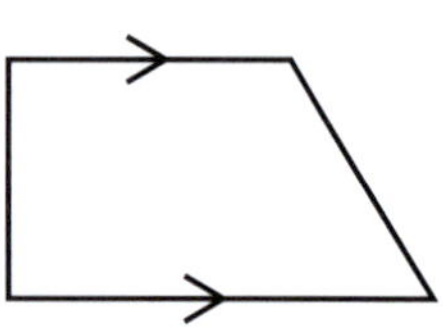

b) 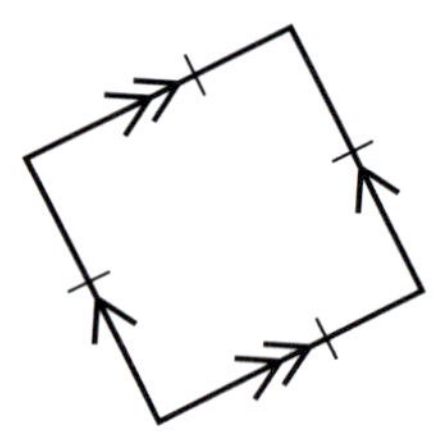

c) 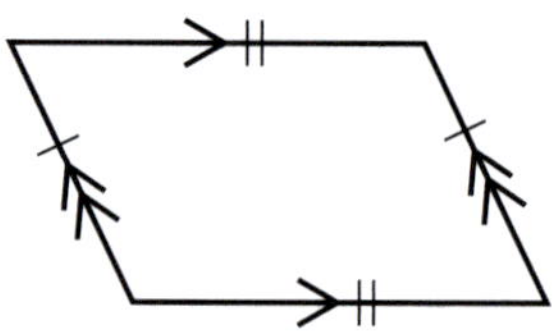

d) 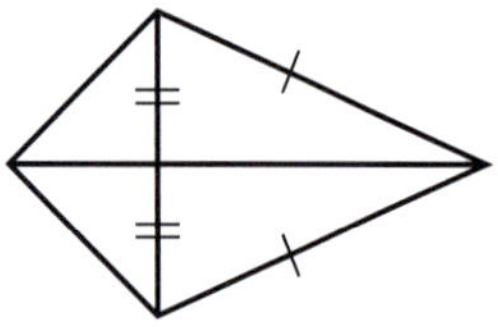

e) 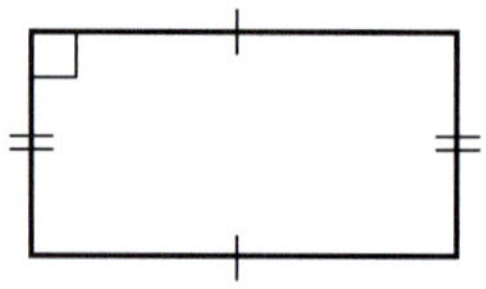

f) 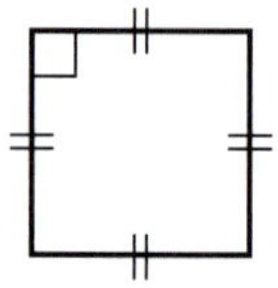

4 Draw an example of each shape and mark it to show its properties.

a) a parallelogram

b) a kite

c) a square

d) a rhombus

5 Find the sizes of the missing angles by calculation:

a)

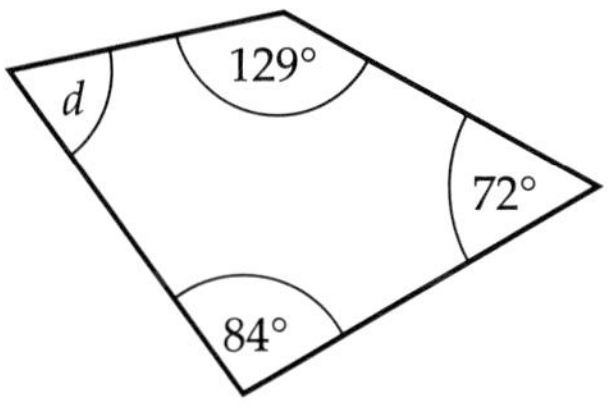

b)

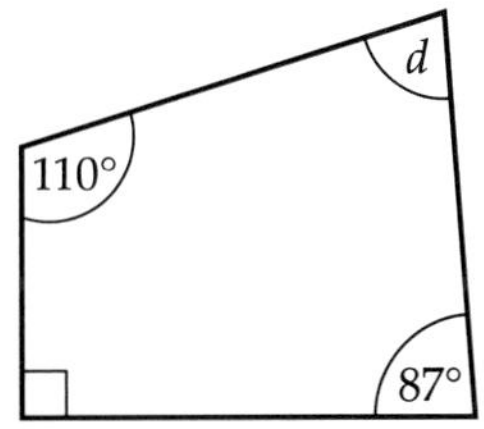

c)

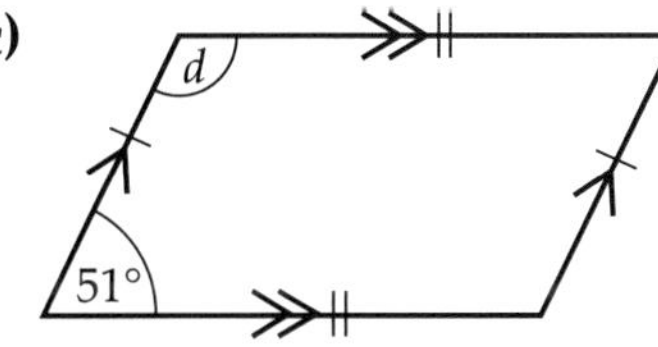

d)

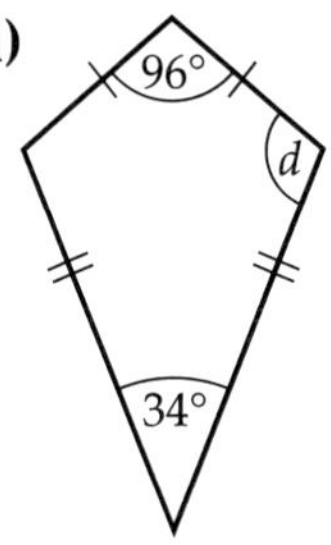

e)

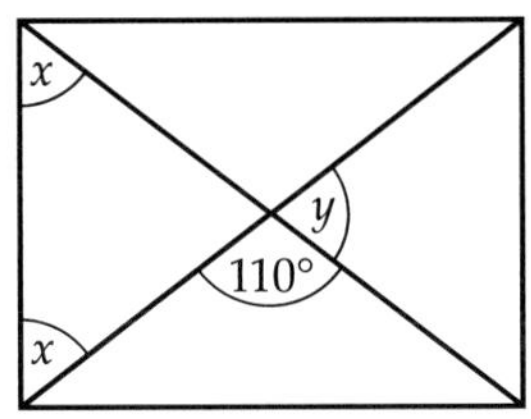

f)

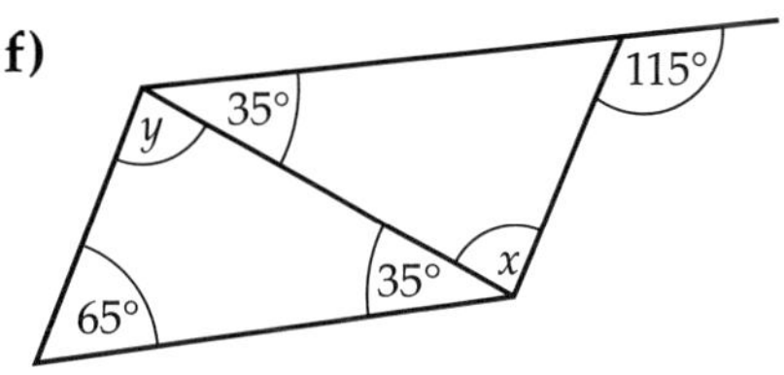

g)

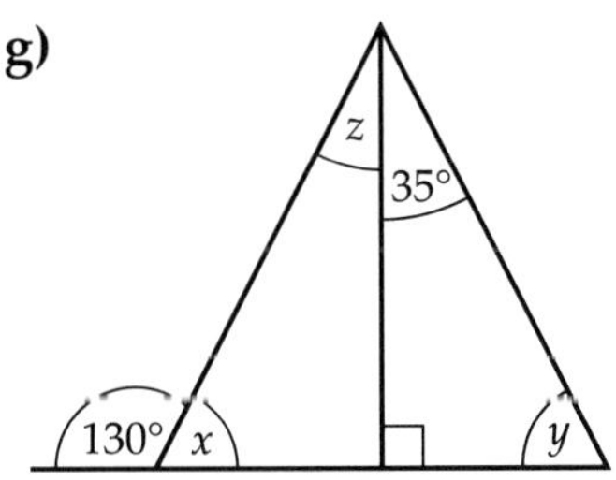

h)

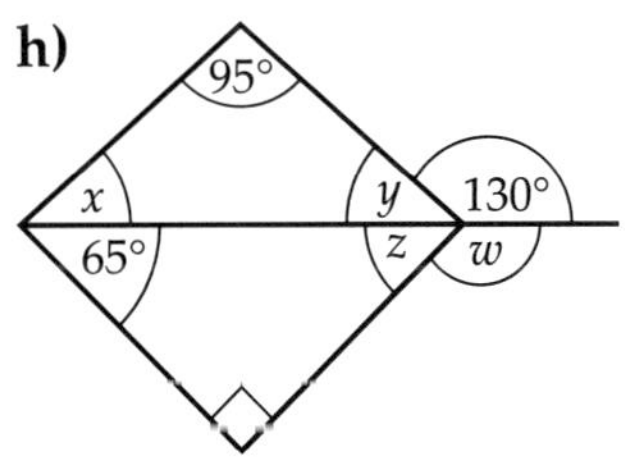

i)

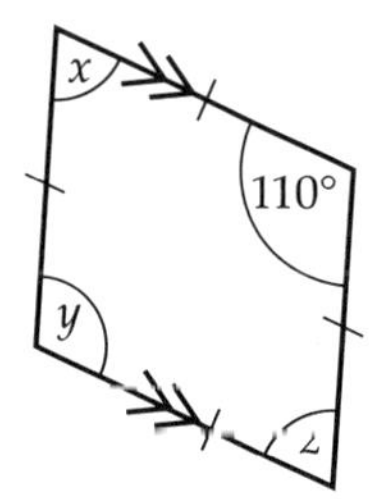

6 Draw a 90° angle and bisect it.

7 Rohini has a length of ribbon 20 cm long.
Draw it and use construction to show where to cut it to make two equal halves.

8 The diagram shows an intersection on a highway.

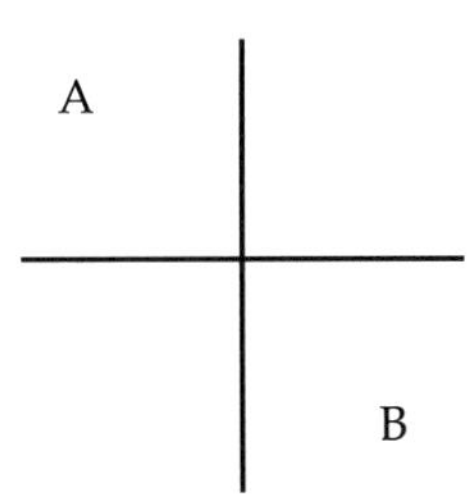

Copy the diagram and show where a telephone line from A to B should be positioned so that it is equidistant from both roads.

Project

1 Muhammed works for Patchworks. He designs decorative shapes for patches on clothing.

a) Copy these shapes onto 1 cm isometric paper.

b) Glue the paper onto card and cut out each shape.

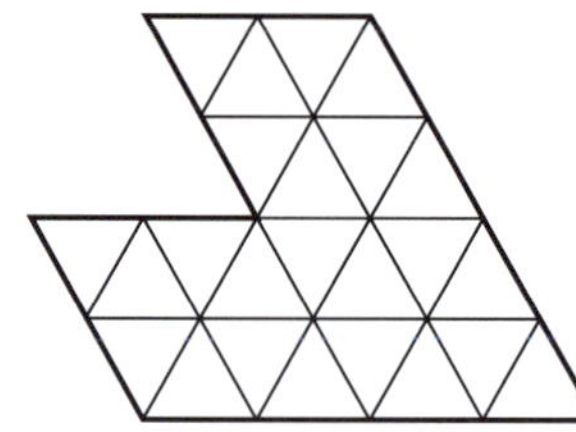

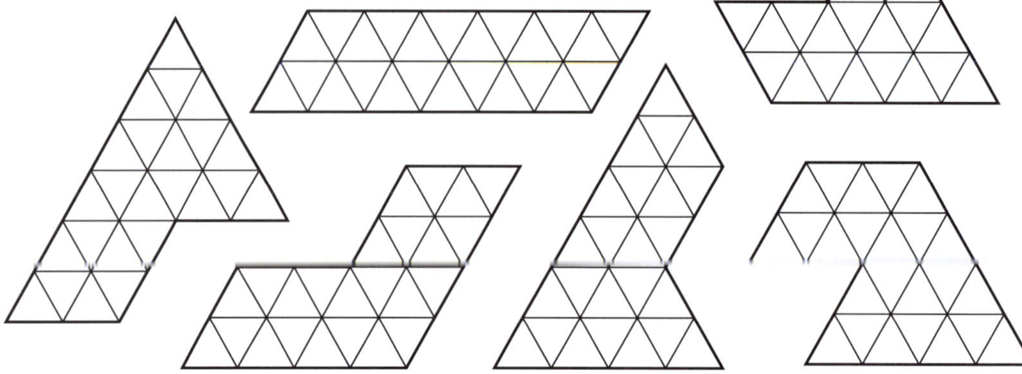

2 Ambika buys these shapes and sews them together to make patches for her jeans.

a) Show how Susan could make this parallelogram patch using four of the shapes.

b) Show how she could make

- a different parallelogram patch using four of the shapes
- another parallelogram patch using five of the shapes.

3 Using some or all of the shapes each time, show how she could make these patches:

a) a rhombus

b) a regular hexagon

c) an equilateral triangle

4 Use the shapes to design some patches of your own.

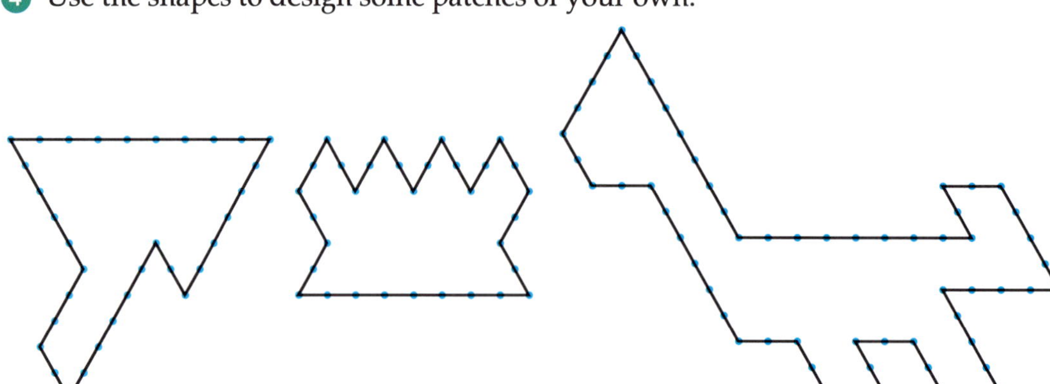

Revise and consolidate

1. Name each solid shown below.

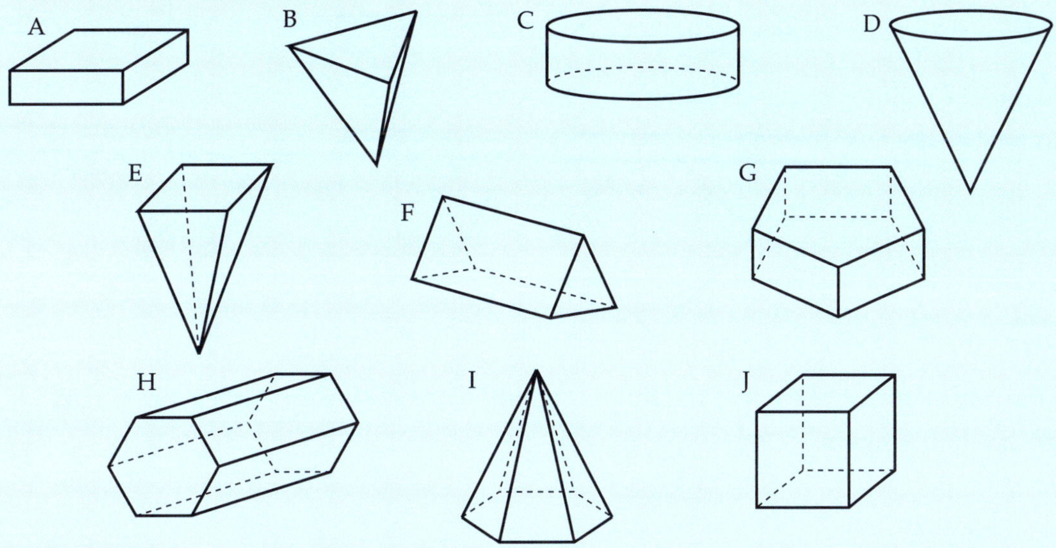

2. Which of the solids above have:

   a) 6 edges

   b) 18 edges

   c) 6 vertices and 9 edges

   d) 7 vertices and 12 edges

   e) 10 vertices?

3. Write the names of the following angles in order from smallest to largest. Say what type of angle each one is.

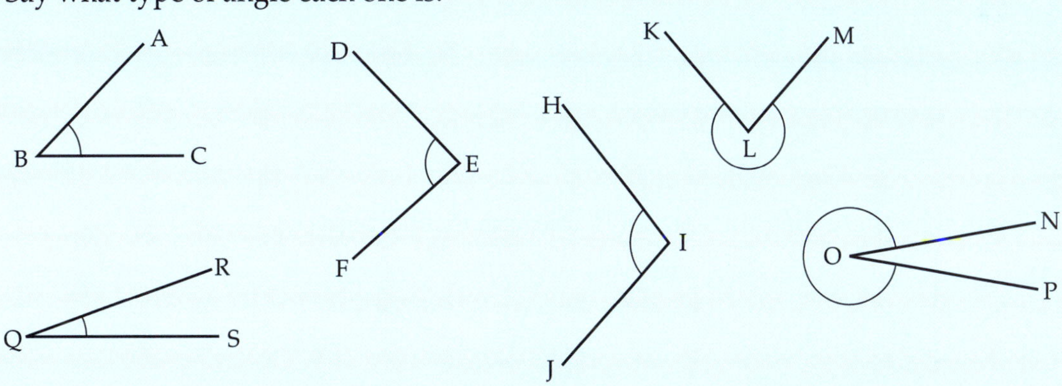

4. Draw the following angles.

   a) $A\hat{B}C = 70°$

   b) $D\hat{E}F = 10°$

   c) $E\hat{F}G = 120°$

   d) $\hat{Q} = 90°$

   e) $X\hat{Y}Z = 195°$

   f) $P\hat{Q}R = 290°$

5 Give the most accurate name for each triangle. Use a ruler and protractor to measure if necessary.

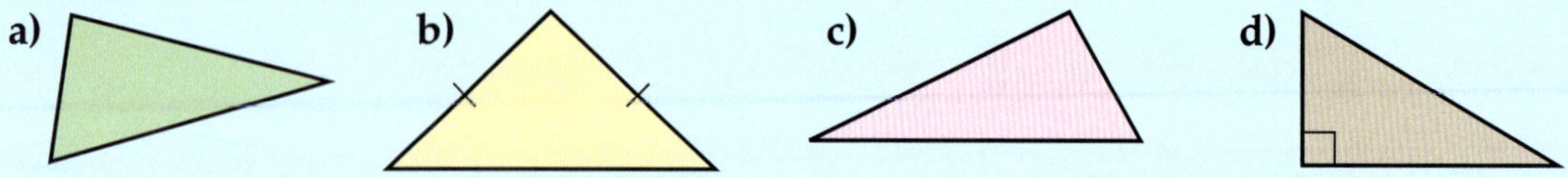

6 Name the following quadrilaterals:

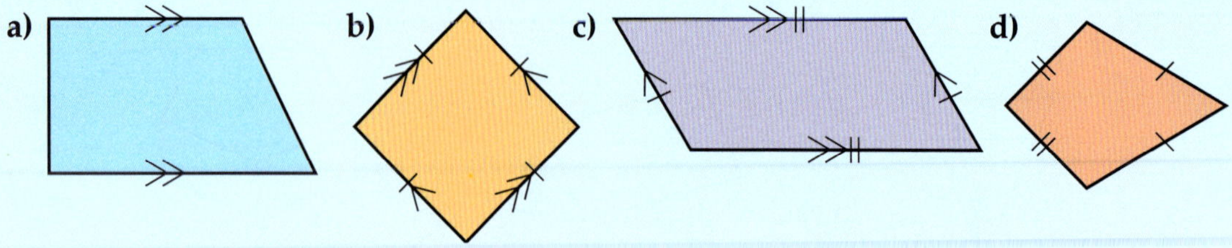

7 Draw an example of each of the following with correct side and other markings:

a) a parallelogram

b) a kite

c) a square

d) an irregular quadrilateral

8 Copy and complete the following table:

Angle	Complement	Supplement
50°		
90°		
60°		
		140°
	2°	
		91°
	45°	

9 Explain why these diagrams are *NOT* mathematically possible.

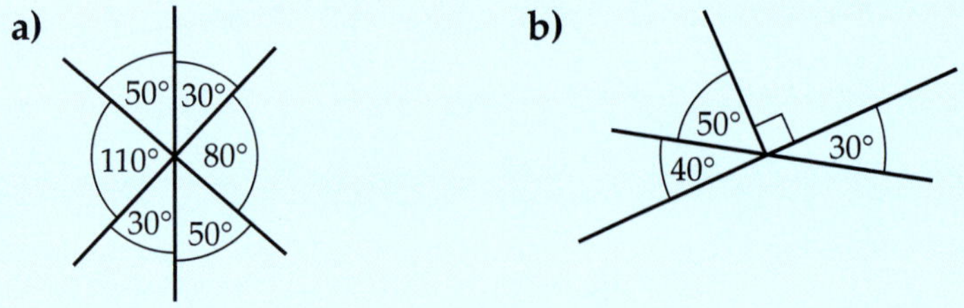

10 Work out the size of the angle named below each of the following.
Write the name of the angle, its size and type.

a)

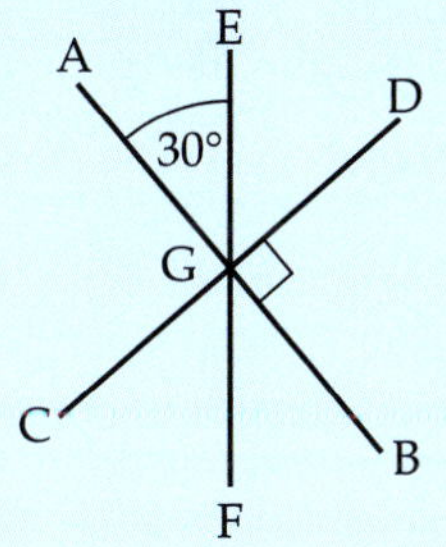

∠AGC

b)

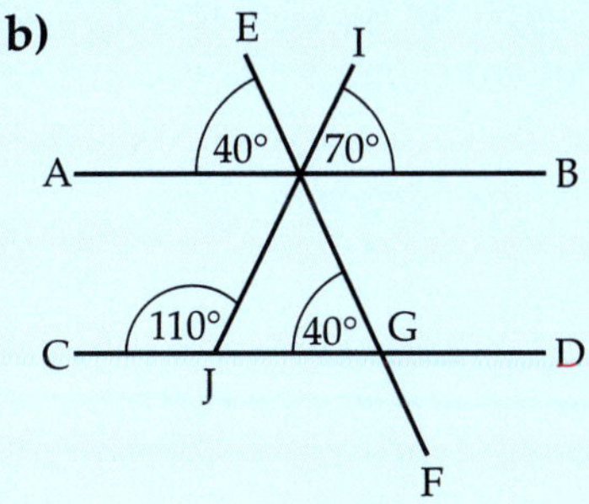

∠EGD

c)

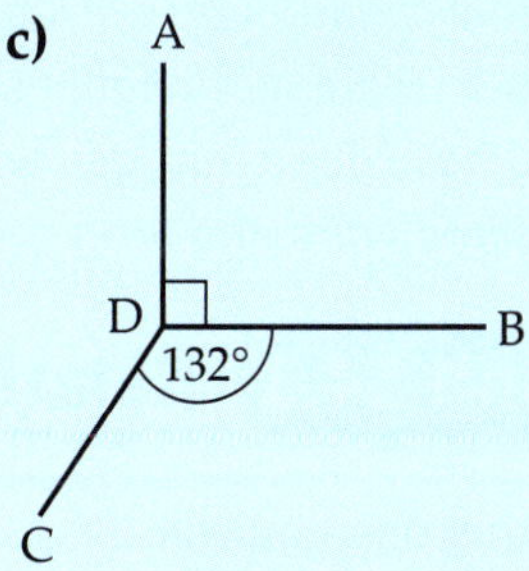

∠ADC

11 Calculate the value of x in each of the following:

a)

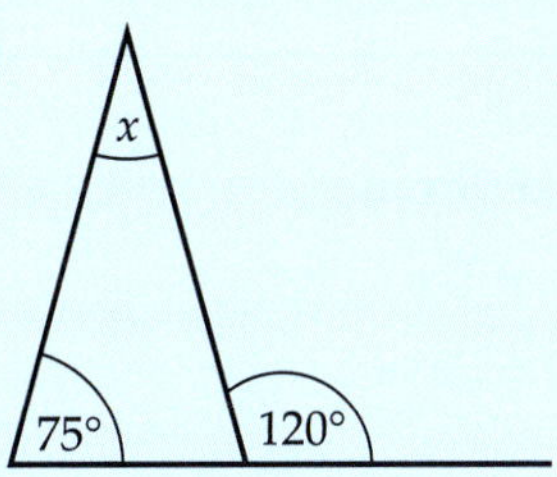

b)

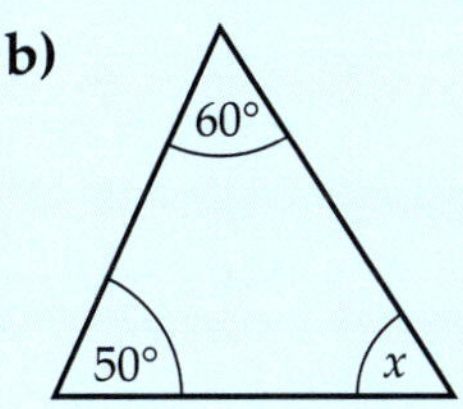

c)

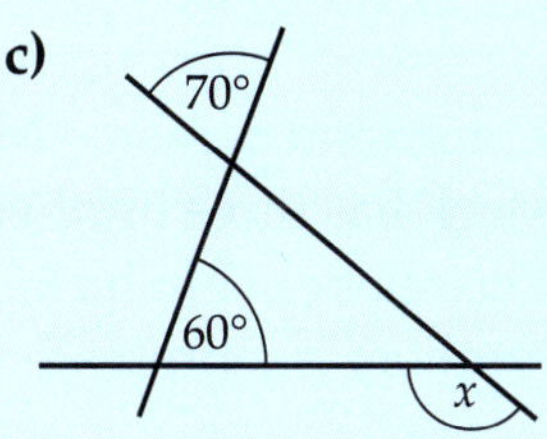

d)

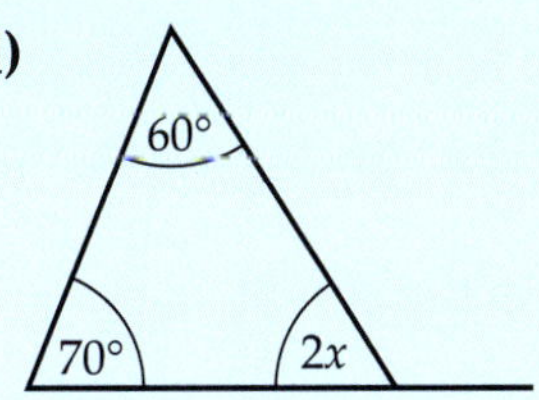

e)

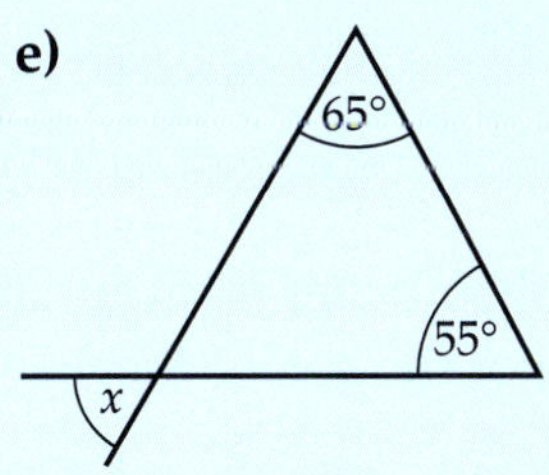

f)

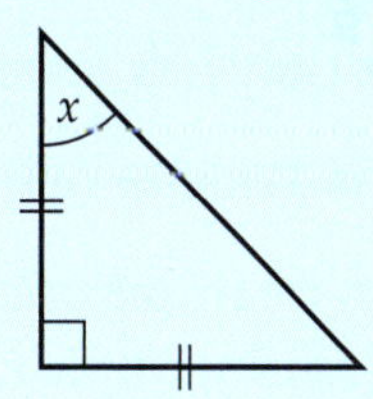

g)

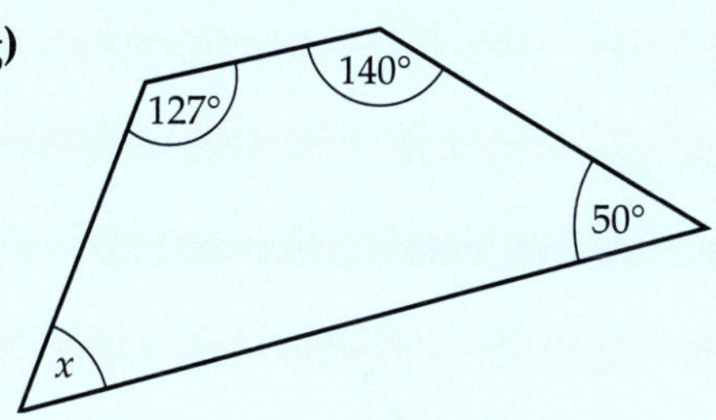

h)

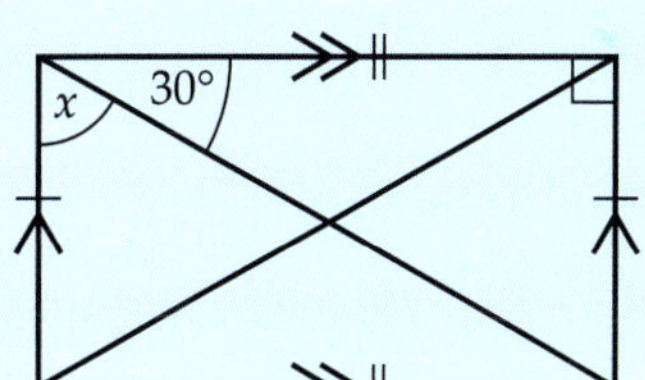

12 The angles of a rectangle are all 90°.
The opposite sides are parallel.
Calculate the value of p and q.
Give a reason for each answer.

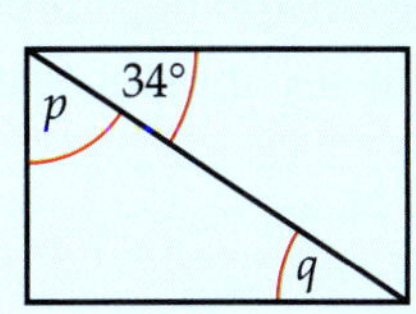

Chapter 14

Relations and relationships

In this chapter you will learn about different types of relations. You will learn how to represent relations using diagrams, mappings and ordered pairs. Once you have done this, you will explore the Cartesian plane and learn how to plot ordered pairs and draw graphs using the Cartesian plane.

Key words
- Relation
- Ordered pair
- Domain
- Range
- Mapping

Unit 14.1 Relations

In this unit you will:

- explore real-life and numerical relations
- represent relations using ordered pairs and arrow diagrams
- identify the domain and range in a relation.

Consider a person – Mark.
Mark has three brothers.
The relationship he has with his brothers is represented here:

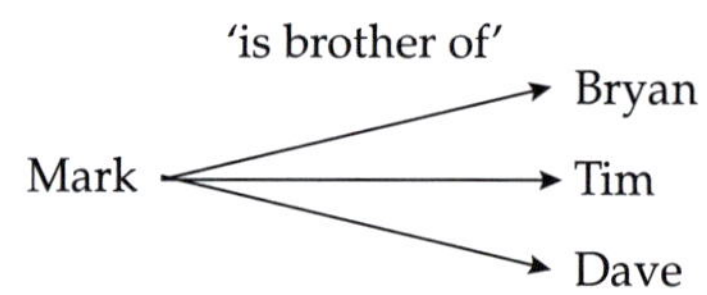

This is a one-to-many relation.

Mark only has one father:

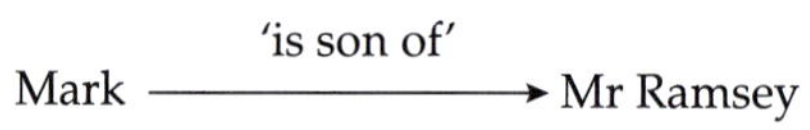

This is a one-to-one relation.

When **relations** are described, their order is important. Mark is the son of Mr Ramsey, but Mr Ramsey is not the son of Mark.

Since the order is important we can write the relation as an **ordered pair**:

(Mark, Mr Ramsey)

Mark's parents also have a relationship with him.

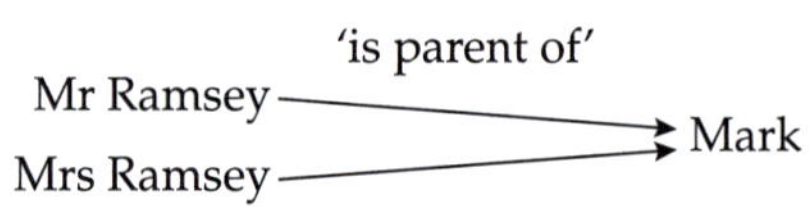

This is a many-to-one relationship.

A relation is a rule by which the elements of one set (the **domain**) are connected with one or more elements of a second set (the **range**).

Worked examples

Write the ordered pairs for the following diagrams:

a.

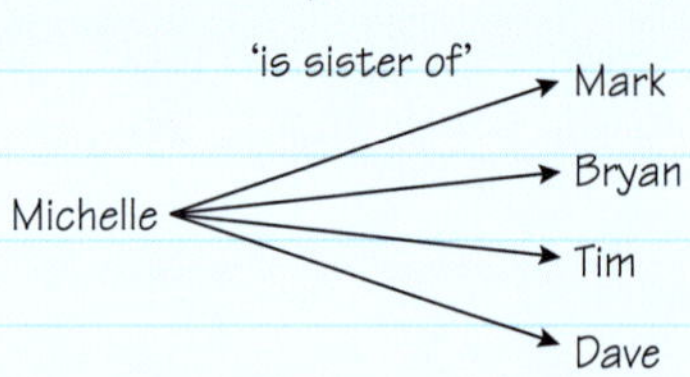

b.

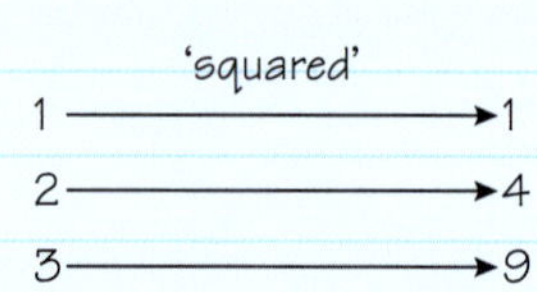

a. (Michelle, Mark), (Michelle, Bryan), (Michelle, Tim), (Michelle, Dave)

b. (1, 1), (2, 4), (3, 9)

Exercise

1 **a)** Draw a diagram to show a relationship in your family.

b) Write down the ordered pairs for your diagram.

2 Draw the diagram which represents the following relation and ordered pairs:

a) relation = 'likes'
(Michelle, blue)
(Michelle, green)
(Michelle, purple)
(Michelle, yellow)
(Sandra, blue)
(Sandra, red)

b) relation = 'dislikes'
(Michelle, brown)
(Michelle, red)
(Michelle, black)
(Ishmael, purple)
(Jonathan, green)

3 For each of the following relations (represented as sets of ordered pairs), draw a diagram and give the domain and range of the relation.

a) (1, 2), (1, 3), (1, 4)

b) (2, 1) , (3, 1), (4, 1)

c) (8, 1), (5, 0), (3, 3), (3, 3)

d) (1, 2), (1, 3), (1, 4), (2, 3), (2, 4)

For a given relation we normally consider several pairs of numbers or elements. For example, for the relation 'is 1 more than' we could have the ordered pairs: (1, 2) (2, 3) (3, 4) (5, 6) (10, 11) and so on.

This can be shown on a diagram using arrows:

$1 \rightarrow 2$	We say 1 is **mapped** onto 2
$2 \rightarrow 3$	We say 2 is mapped onto 3
$3 \rightarrow 4$	We say 3 is mapped onto 4
$5 \rightarrow 6$	We say 5 is mapped onto 6
$10 \rightarrow 11$	We say 10 is mapped onto 11

Notice that we have two sets of numbers: the **domain**, those on the left, and **range**, those on the right. In an ordered pair, the domain is written first, the range second.

Exercise

1 List the members of the domain and range in the following:

a) (3, 7) (4, 9), (6, 13)

b) (2, 8), (3, 27), (4, 64)

c) $(1, \frac{1}{2})$, (2, 2), (6, 18)

2 In relation 'is double', the following domains are given:

2, 5, 10, 12, 200

List these as ordered pairs by filling in the range.

3 State one possible relation in the following cases:

a) Domain 2, 4, 6 range 1, 2, 3

b) Domain 1, 2, 3 range 0, 1, 2

c) Domain 5, 10, 15 range 25, 100, 225

d) Domain 3, 6, 9 range 18, 18, 18

e) Domain 100, 120, 140 range 10, 12, 14

Unit 14.2 Different types of relations

Key words
One-to-one
Many-to-one
One-to-many
Many-to-many

In this unit you will:

- identify different types of relations
- learn to represent relations using arrow diagrams.

- Mark is the only son of Mr Ramsey. They have a **one-to-one** relationship.
  Mark is the only son Mr Ramsey has.
  Mr Ramsey is the only father Mark has.

4 is the only number that is half as much as 8.
8 is the only number that is twice as much as 4.

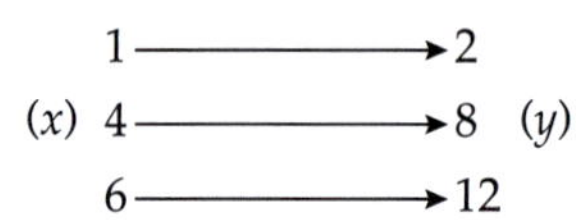

These are one-to-one relations. In any ordered pair (x, y), one value of x maps onto one, and only one, value of y.

- Ms Kennedy has many students.
  Each student in her class has only one teacher.

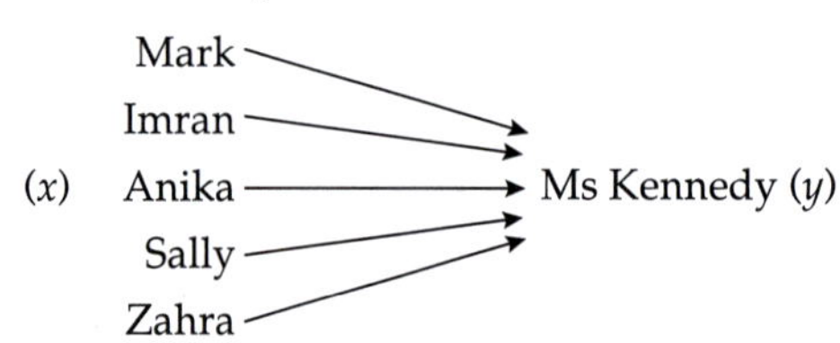

September, April, June and November all have 30 days.
No other months have 30 days.

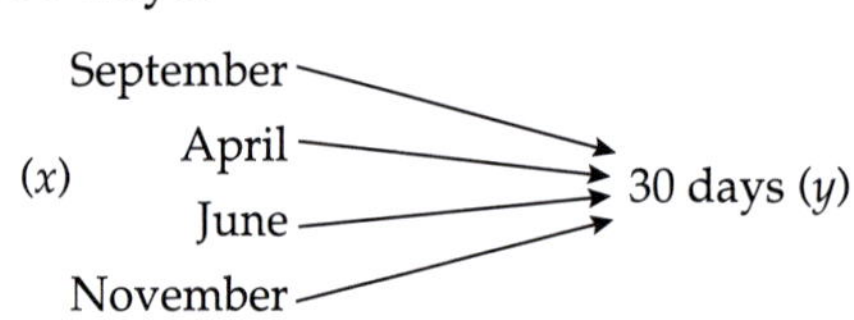

These relations are **many-to-one** relations because many items, or values, map onto one item or value. In any ordered pair (x, y) there are many values for x, but only one value for y.

- Afiya is one student who studies many subjects.
  This is a **one-to-many** relation. In any ordered pair (x, y) there is one value for x, but many values for y.

- In a **many-to-many** relation some x-values map to many y-values and some y-values are mapped to by many x-values. The arrow diagrams can become quite complex.

Exercise

1 For each arrow diagram, a relation is shown. Suggest what the relation may be.

a) 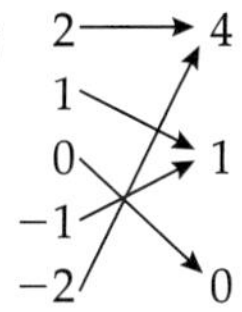

b) 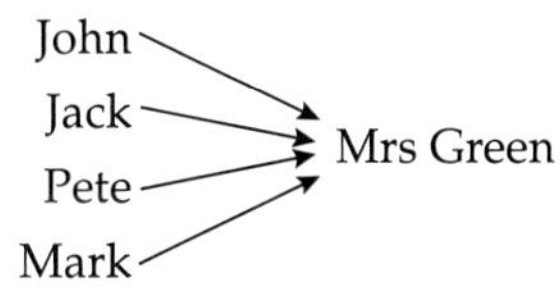

c) 

d) 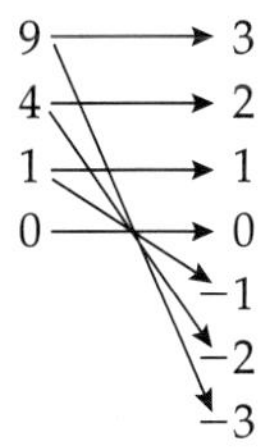

e) 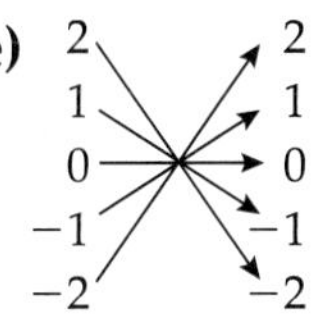

f) 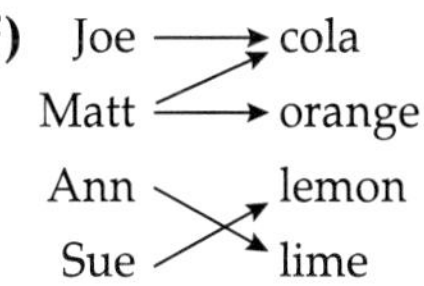

2 Draw arrow diagrams to represent the following. Explain the type of relation to which each may belong.

a) Ishmael and Muhammed are both sons of Mr and Mrs Khan.

b) Mrs Prevatt has three daughters: Susan, Millie and Jess.

c) Joe and Matt play soccer. Sue and Lisa play hockey. Joe plays hockey as well.

d) Peter and Mary are brother and sister.

e) Shaiwa and Kelly Anne both like blue, Taneisha likes red and Paula likes green.

f) Domain: 3, 5, 7, 9; range: odd, prime, multiple of 3

3 What type of relation is indicated by each set of ordered pairs?

a) (2, 4), (4, 8), (6, 12)

b) (2, 4), (2, 6), (2, 8)

c) (4, 2), (6, 2), (8, 2)

d) (2, 3), (3, 4), (4, 2), (2, 4)

Investigation

What relations exist between people?
List ten different types of relations that you can think of.
Represent these using arrow diagrams.
Exchange diagrams with a partner.
Try to work out the relations shown in each arrow diagram and identify the type of relation shown.

Consider the numbers from 1 to 25.
Draw diagrams to illustrate at least five relations using this set of numbers.
State each relation in words and say what type it is.

Unit 14.3 Mappings

Key words
Mapping

In this unit you will:

- be able to use and identify a mapping
- recognise some of the properties of a mapping diagram.

A **mapping** is a rule which connects one set of numbers to another set of numbers.

A mapping looks like: $x \rightarrow x + 4$

We can display a mapping in a mapping diagram:

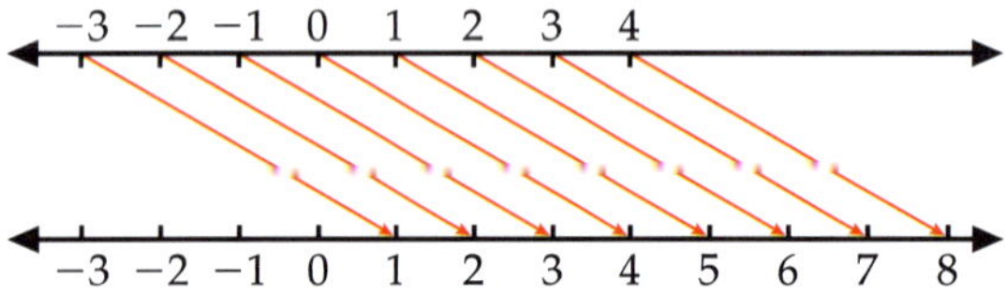

Each number on the top number line is mapped to the number four more than it on the bottom line.

Or we can display it in the following way:

$-2 \rightarrow 2$
$-1 \rightarrow 3$
$0 \rightarrow 4$
$1 \rightarrow 5$

Worked examples

1. Copy and complete the following pair of number lines, for x from 0 to 6, given the mapping: $x \rightarrow 2x - 5$

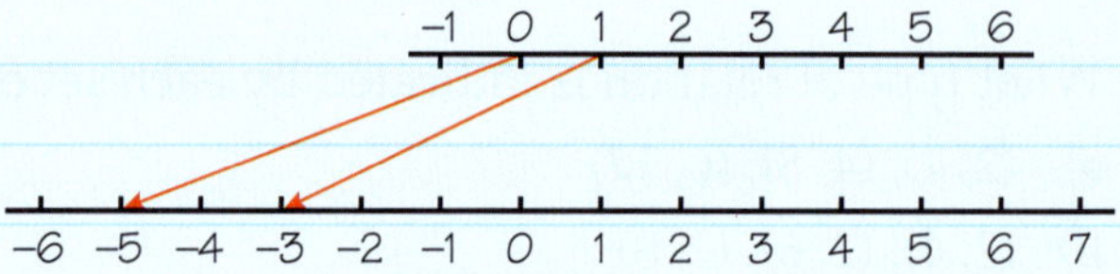

2. Identify the mapping for these pairs of numbers:

$0 \rightarrow 0$
$1 \rightarrow 2$
$2 \rightarrow 4$
$3 \rightarrow 6$

1.

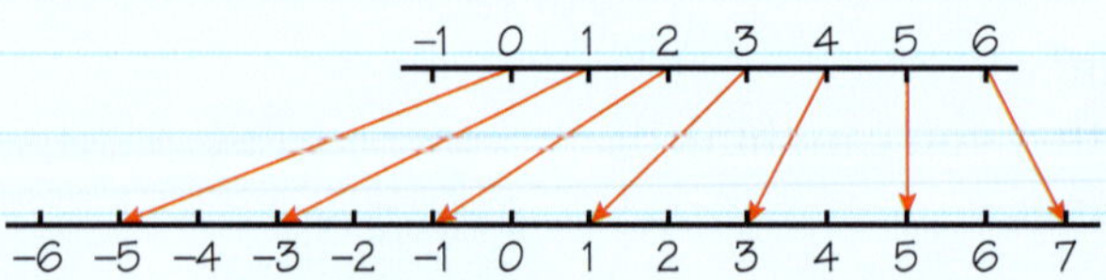

To complete the diagram of the mapping we need to show what each number on the top line will become after we have multiplied by 2 and subtracted 5.

2. $x \rightarrow 2x$

Each time the number is being doubled.

Exercise

1 Draw a mapping diagram for each of the following mappings. For each part of the question, draw a pair of number lines from −10 to +10.

a) $x \rightarrow x + 10$ **b)** $x \rightarrow x - 4$ **c)** $x \rightarrow 3x$ **d)** $x \rightarrow \frac{x}{2}$

2 Given the mapping fill in the following:

a) $x \rightarrow 5x - 4$

$1 \rightarrow \square$
$2 \rightarrow \square$
$3 \rightarrow \square$
$4 \rightarrow \square$

b) $x \rightarrow \frac{x}{5} + 7$

$0 \rightarrow \square$
$5 \rightarrow \square$
$10 \rightarrow \square$
$15 \rightarrow \square$

c) $x \rightarrow \frac{x-1}{11}$

$12 \rightarrow \square$
$23 \rightarrow \square$
$34 \rightarrow \square$
$45 \rightarrow \square$

d) $x \rightarrow \frac{2x}{5}$

$0 \rightarrow \square$
$5 \rightarrow \square$
$10 \rightarrow \square$
$15 \rightarrow \square$

3 Identify the mapping for each of the following pairs of number lines:

a)

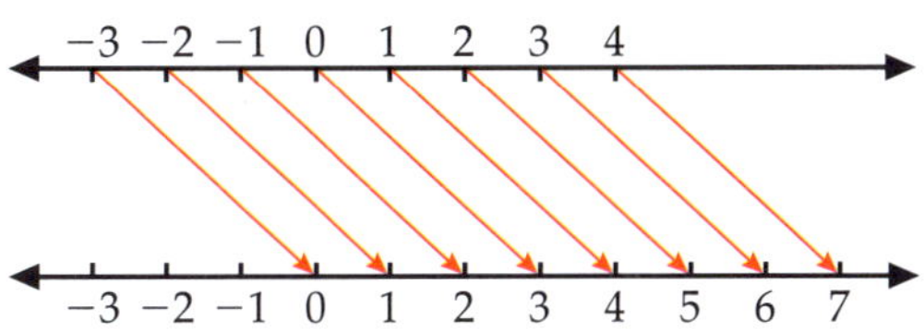

b)

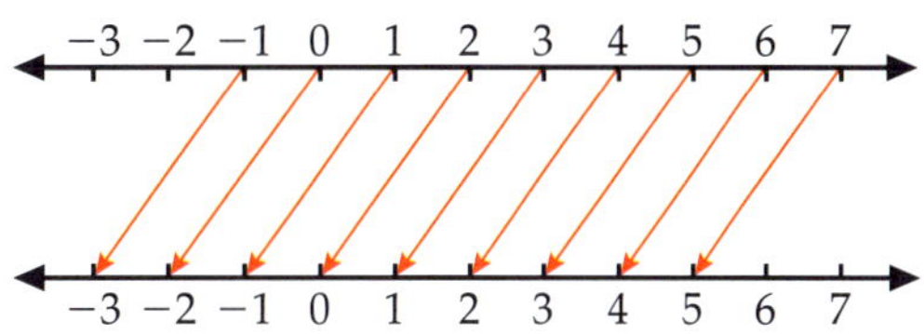

4 Identify the mapping for each of the following sets of numbers:

a) $5 \rightarrow 11$
$6 \rightarrow 12$
$7 \rightarrow 13$

b) $1 \rightarrow 0$
$2 \rightarrow 1$
$3 \rightarrow 2$

c) $1 \rightarrow 5$
$2 \rightarrow 10$
$3 \rightarrow 15$

d) $30 \rightarrow 5$
$60 \rightarrow 10$
$120 \rightarrow 20$

5 **a)** If you mapped a positive number using the mappings below, would you end up with a smaller or larger number?

i) $x \rightarrow x + 5$ **ii)** $x \rightarrow \frac{x}{10}$

iii) $x \rightarrow x - 4$ **iv)** $x \rightarrow 8x$

Try each of the mappings out on a particular number.

b) Describe any pattern that you notice.

6 **a)** Sketch mapping diagrams for the following mappings.

i) $x \rightarrow x + 4$ **ii)** $x \rightarrow x + 1$ **iii)** $x \rightarrow x + 12$

b) What do you notice about the arrows you have drawn on your mapping diagrams?

c) Explain your answer to part **b)**.

7 Draw mapping diagrams of the form:

$x \rightarrow x - \square$ where $\square$ can be replaced by any number.

Describe any patterns you notice on your diagrams and explain why this is the case.

8 In a car manual the following information is given to show how far you can drive with a given number of litres of petrol.

a) What is the mapping that connects the number of litres of petrol to the number of miles?

b) If you had 5 litres of petrol how far could you drive?

c) If you had 40 litres of petrol how far could you drive?

d) If you drove 42 miles how many litres of petrol did you use?

Litres		**No. of miles**
1	→	7
10	→	70
15	→	105
20	→	140
50	→	350

Unit 14.4 Coordinates on a grid

In this unit you will:

- learn about and use coordinates
- plot points on a grid using coordinates.

Key words
Points
Axis/axes
Coordinates
Grid

You have already worked with **points** in one dimension on number lines. For example, the position of point A on this number line is 3. The position of point B is 4.5. The position of the point is given from 0 which is a fixed point.

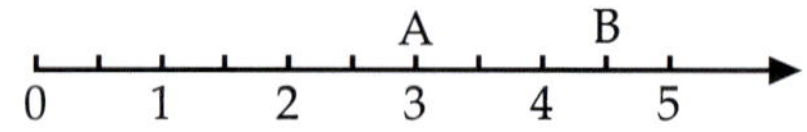

We can also show points in two dimensions. To do this, we use two number lines, one horizontal and one vertical. You have used a system like this to draw graphs. The number lines in such cases are called **axes**. The horizontal **axis** is called the x-axis and the vertical one is called the y-axis.

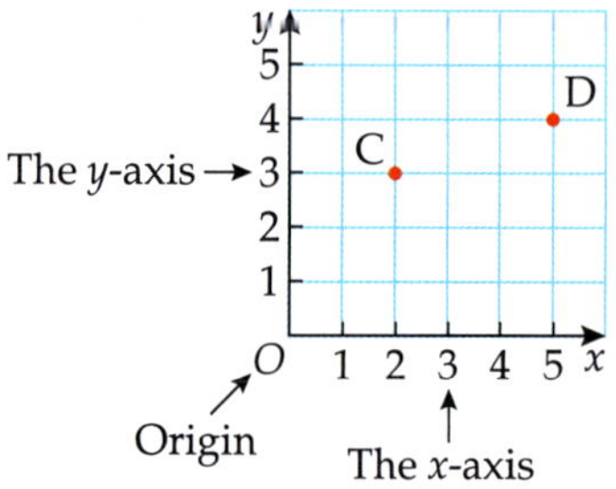

In the example above we have to give two numbers, or **coordinates**, to give the position of the points. The x-coordinate tells you how far along the x-axis a point is, while the y-coordinate tells you how far up the y-axis it is. The x- and y-coordinates are written together in brackets. The x-coordinate is always given first. For example, the points on the grid above are at (2, 3) and (5, 4).

One way of remembering how to write coordinates is to remember that x comes before y in the alphabet.

The process of placing dots at given positions on a **grid** is called plotting points.

When you use a road map or a map in an atlas you sometimes have to locate places on the map. Most atlases provide an index at the back which gives the names of places, the pages on which they can be found and coordinates (usually in letters and numbers) to help you locate the exact position of places on the map. On the map below, the letters at the top and bottom and the numbers along the sides together form a square in which we can locate places. For example, Pebbles Hill is found in square B2. The letter B and the number 2 together form a pair of coordinates. These are general positions rather than exact points.

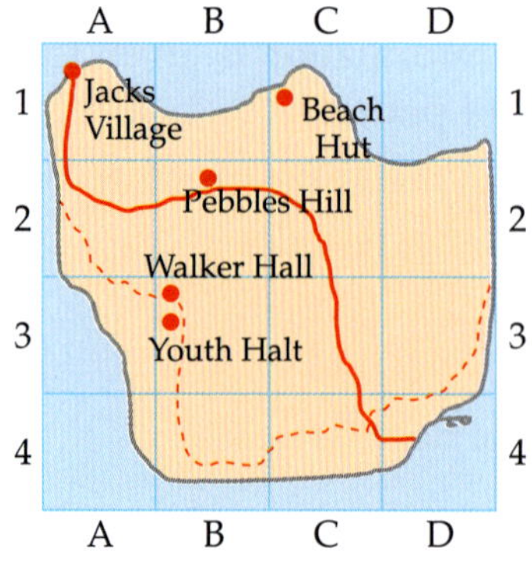

Worked examples

1. Plot the following points on a grid.

   A = (2, 1) B = (1, 2) C = (0, 5) D = (4, 4)

2. What are the coordinates of points M, N and O?

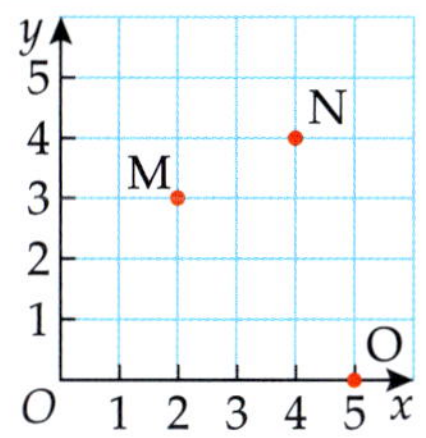

1. Start by drawing and labelling the axes.

   Find and plot each point.

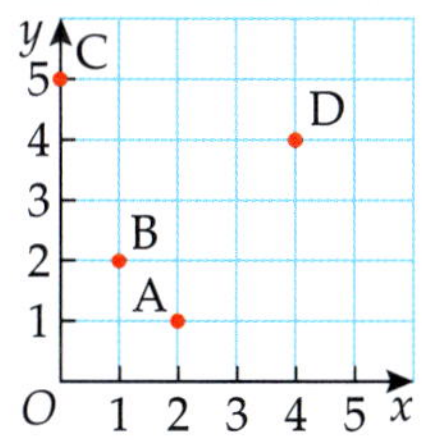

2. M = (2, 3); N = (4, 4); O = (5, 0)

Exercise

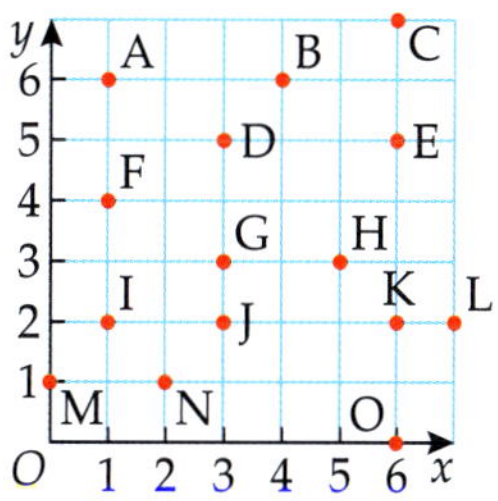

1 Write the coordinates of the following points from the grid:

a) A	**b)** B	**c)** C
d) D	**e)** E	**f)** F
g) G	**h)** H	**i)** I
j) J	**k)** K	**l)** L
m) M	**n)** N	**o)** O

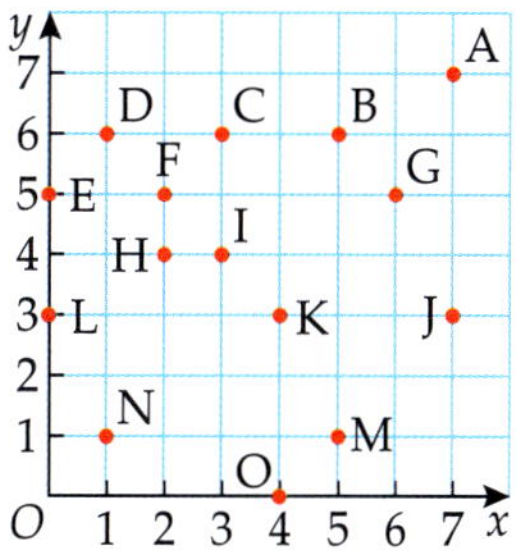

2 Write the capital letter used to label each of the following points on this set of axes.

a) (1, 1)	**b)** (7, 7)	**c)** (3, 4)
d) (6, 5)	**e)** (4, 3)	**f)** (5, 6)
g) (0, 5)	**h)** (2, 4)	**i)** (7, 3)
j) (3, 6)	**k)** (1, 6)	**l)** (0, 3)
m) (4, 0)	**n)** (2, 5)	**o)** (5, 1)

3 Write the coordinates of the places on the map on page 148.

Unit 14.5 The Cartesian plane

Key words
Cartesian plane
Quadrant

In this unit you will:

- learn about the Cartesian plane and how it is used
- plot and graph patterns on the Cartesian plane.

The grid system most commonly used in mathematics was invented in 1619 by a French mathematician named René Descartes. This grid system can be used to locate single points on a grid exactly, even when the coordinates of those points are fractions. This gives it a great advantage over systems such as those used in street directories which only tell you the square of the map to look in to find a particular feature.

The diagram below shows you how the **Cartesian plane** (named after Descartes) works. This year you are only going to work with the first **quadrant** of the plane.

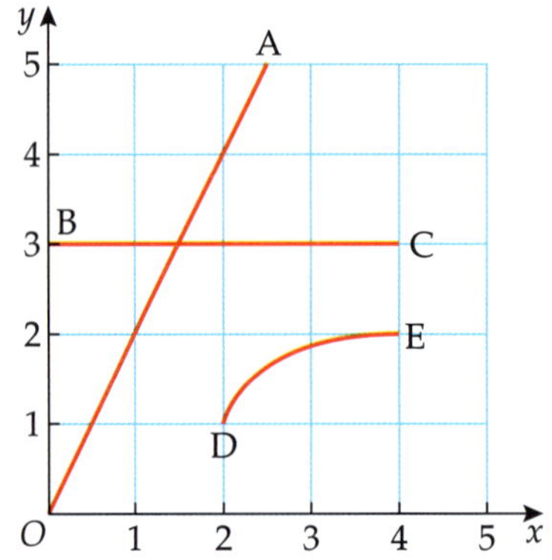

When points on the Cartesian plane are joined, as in OA, BC and DE, lines are formed. The shortest distance between two points is a straight line. Points are usually joined in the order in which they are given.

The ordered pairs or coordinates sometimes show a pattern which can be written as an equation. For example:

- Some coordinates on OA are (0, 0), (1, 2) and (2, 4). Can you see that the y-coordinate is double the x-coordinate in each case? We can write this as $y = 2x$.
- Some coordinates on BC are (0, 3), (1, 3) and (2, 3). In this case, the y-coordinate is always 3. We can say $y = 3$.

Exercise

For questions 1 to 3, work on a 10 × 10 cm grid.

1 **a)** Plot the points A(1, 3), B(2, 6) and C(3, 9), then join them in the order given.
b) Extend the line segment and write three other sets of ordered pairs from this line.
c) Write an equation for the pattern of the ordered pairs.

2 **a)** Plot the points (5, 1), (5, 3), (5, 5) and connect these points.
b) Write an equation to describe this line.

3 **a)** Plot the points O(0, 0), P(4, 0), Q(4, 4) and R(0, 4).

b) Join the points and name the shape.

Join points in the order given.

4 The following sets of points may be joined in order to form aquatic creatures.

a) Draw a Cartesian grid to 18 on the x-axis and to 11 on the y-axis. Plot:

(5, 6) (2, 6) (3, 5) (5, 4) (8, 4) (10, 2) (10, 4) (15, 5) (17, 4) (16, 6) (17, 8) (15, 7) (10, 8) (10, 10) (8, 8) (5, 8) (3, 7) (2, 6)

b) Draw an x-axis to 17 and a y-axis to 8. Plot:

(11, 1) (12, 1) (13, 2) (13, 3) (16, 3) (14, 4) (12, 6) (10, 7) (8, 7) (6, 6) (4, 4) (3, 4) (3, 5) (2, 5) (1, 4) (1, 3) (5, 3) (5, 2) (4, 1) (5, 1) (6, 2) (6, 3) (12, 3) (12, 2) (11, 1)

Looking back

1 **a)** Plot the points: (10, 2), (9, 4), (7, 5) (2, 10), (6, 1).

b) Plot the points: (1, 2.4), (9.6, 4), (7.5, 0.5), (2, 10), (0.6, 1). Label the x-axis and the y-axis.

2 Plot the points: (0, 5), (3, 0), (7, 0), (10, 5), (7, 10), (3, 10), (0, 5) and join them. What is the name of the shape drawn?

3 **a)** Plot any three points on the x-axis and write the ordered pairs.

b) Plot any three points on the y-axis and write the ordered pairs.

c) Complete the following statements:

On the x-axis $y =$ ______

On the y-axis $x =$ ______

The coordinates of the origin are ______ .

4 Consider these equations which describe lines on a plane:

a) $y = 6$ **b)** $x = 8$ **c)** $y = x + 4$

d) $y = 2x + 1$ **e)** $y = 3x - 1$

When $x = 1$ and 3, find the value of y for each line except **b)**. Use these values to write two sets of ordered pairs.

Plot the points and draw all the lines.

5 Copy the line and shapes in the grid. Plot the following points and join them in the order in which they are given. A treasure is hidden at the end. Reach for it.

B(4, 3)	J(1, 1)
C(5, 3)	K(1.5, 1)
D(5, 4)	L(1, 0)
E(4, 4)	M(4, 0)
F(3, 5)	N(3.5, 1)
G(1, 5)	O(4, 1.5)
H(0.5, 2)	A(2, 2)

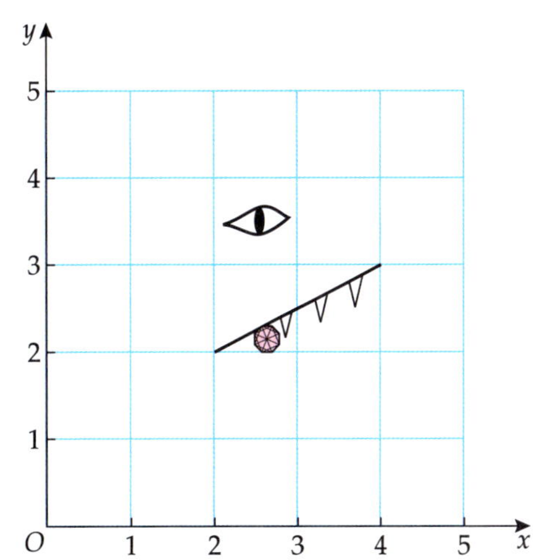

Project

Piet Mondrian (1872–1944) was a Dutch abstract artist. Many of his works are based on a series of intersecting lines and so can be described using the Cartesian coordinate system.

For this painting, we might imagine that the left-hand edge occurs at the *y*-axis and that the bottom is on the *x*-axis. We could add values to the axes at appropriate points, as we have done, and can use this to copy the grid lines and then to colour the painting.

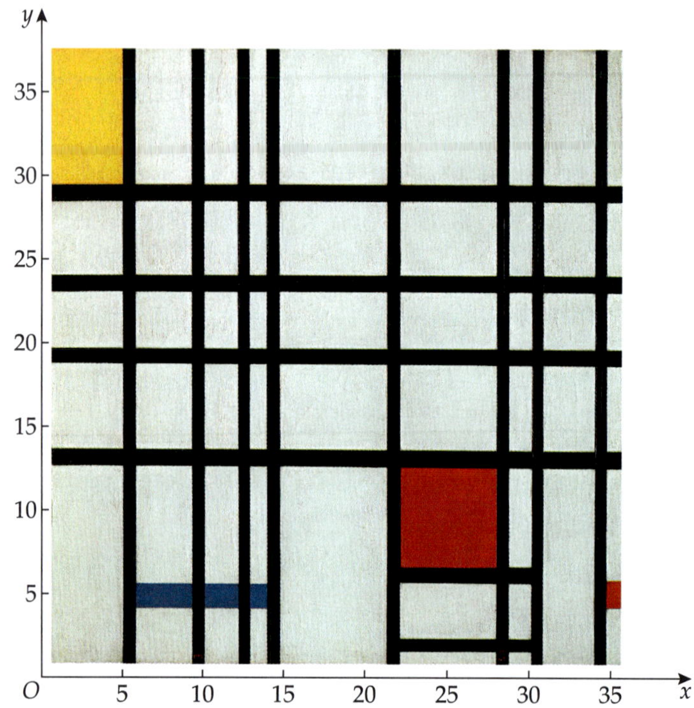

1 Use the method above to make a large copy of Mondrian's painting.

2 **a)** Draw and colour a Mondrian style painting of your own. Show the grid that you use.

b) Describe your painting to a partner in words using grid references (for example (2, 5)).

Key words
Data
Survey
Questionnaire
Interview
Tally
Primary sources
Secondary sources

Chapter 15

Working with data

We have access to a great deal of information from books, the media, the Internet and other people. In this chapter you will learn more about identifying sources of information, collecting, sorting and representing information. You will also learn how to read and draw different kinds of graphs.

Unit 15.1 Collecting and recording data

In this unit you will:

- collect information on a topic
- record information collected.
- draw a tally chart

Data is another word for information. The branch of mathematics called statistics is about collecting data and showing the data in such a way that it can be easily understood. This may be in the form of a table, chart or diagram.

Data may be collected from **primary sources** by carrying out a **survey**, where you can observe what is happening or ask questions; or it may be collected from **secondary sources** such as newspapers, magazines, encyclopaedias or the Internet. Sometimes data has to be collected by conducting an experiment.

If all of the data cannot be collected by observation, it will be necessary to get information from other people. In an **interview**, you will ask questions and the people will answer at the same time. If you prefer, the questions may be written down as a **questionnaire** and distributed.

The data that is collected must be recorded so that it can be used. A simple way of recording data is to use a table. The table may contain **tally** marks where each one represents a single occurrence of the event. An event which occurs once will have one tally |, an event which occurs three times will have three tallies ||| and an event which occurs four times will have four tallies ||||. However, for the fifth tally, a line is drawn through the previous four ~~||||~~. This makes it easy to count the tallies made in groups of five.

Worked examples

1. Which of the following data can be best collected by:
   - observation
   - an interview
   - an experiment?

   a. the number of Toyotas in a car park
   b. the heights of seedlings after 1 week
   c. the favourite cartoon of the students in the class
   d. the number of students in the class who wear glasses
   e. the number of hours spent doing homework

2. The table below shows the number of library books borrowed by three students. How many books did each student borrow?

Name	No. of books
Sherry-Ann	~~||||~~ |||
Omar	||
Adam	~~||||~~ ~~||||~~

1. a. Observation: the cars will be observed and the make noted
   b. Experiment: the heights will be measured after one week and the heights recorded
   c. Interview: the students will be asked what their favourite cartoon is
   d. Observation: the students wearing glasses may be asked to stand and are counted
   e. Interview: each student will be asked how many hours are spent doing homework
2. Sherry-Ann borrowed 8 books, Omar borrowed 2 books and Adam borrowed 10 books.

Exercise

1. For each of the following questions, decide whether doing a survey, an experiment or collecting data from a secondary source would be the best way of getting data.
   a) Are house rents in your parish higher than those in another parish on the island?
   b) Do people with longer legs run faster?
   c) How many times per month does the average Form 1 student eat at a restaurant?
   d) What is the number of goals scored by the school football team in a season?
   e) What is the favourite television cartoon of the pupils in the class?
   f) How many peas are there in a pod?

2. Juan is carrying out an experiment to test if a standard six-sided dice is fair. He throws the dice ten times and gets these results.

Number on dice	1	2	3	4	5	6
Frequency	2	1	4	0	2	1

   Juan says that because he did not get any 4s, his dice was not fair.
   What was wrong with his experiment?

3. Explain why these surveys could lead to incorrect or biased results:
   a) Asking: 'What is your favourite sport?' outside a cricket ground.
   b) Asking: 'How many books do you read each week?' in the school library.
   c) Asking: 'Do you buy food from the sweetshop?' in the queue at the school sweetshop.

4. Which units would be sensible to use if you were carrying out an experiment or survey on the following topics?
   a) Are Form 1 students shorter than Form 2 students?
   b) How far do you live from school?
   c) How long do you spend on homework each day?
   d) Is there a connection between the age of a pupil and the size of his or her school bag?
   e) Are shoe size and handspan related?
   f) Is age connected to the amount of sleep children need?

Key words
Frequency table
Pictograph
Scale

Unit 15.2 Frequency tables and pictographs

In this unit you will:

- develop frequency tables
- draw and use pictographs.

The data collected from surveys and investigations must be presented in such a way that it can easily be used. In Unit 15.1, we looked at a table of tally marks. Data may also be shown in a **frequency table**, which shows the number of times each event occurs.

Side a coin lands on	Tally marks	Frequency
Heads	𝍸 𝍸 𝍸	15
Tails	𝍸 𝍸 ||	12

However, pictures and diagrams are a more interesting way of presenting the data. In addition, it is easier to glance at a diagram and see the trends. The **pictograph** is a diagram which uses a picture as a symbol to represent each occurrence of an event. If there is a large amount of data, a **scale** must be used where each picture may be used to represent a number of occurrences.

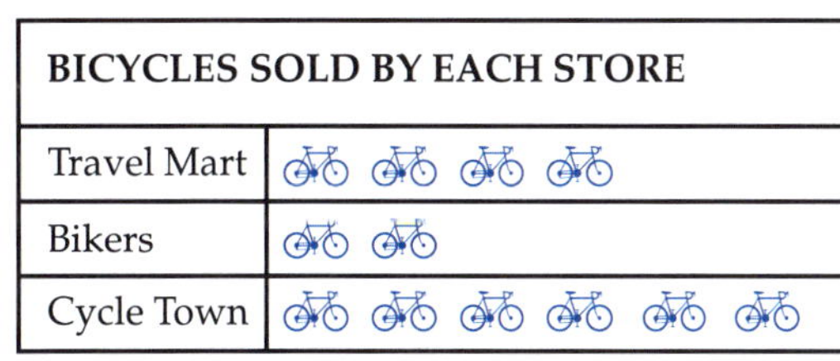

Worked examples

The pictograph below shows how many children attended the cinema in a certain week during the holidays.

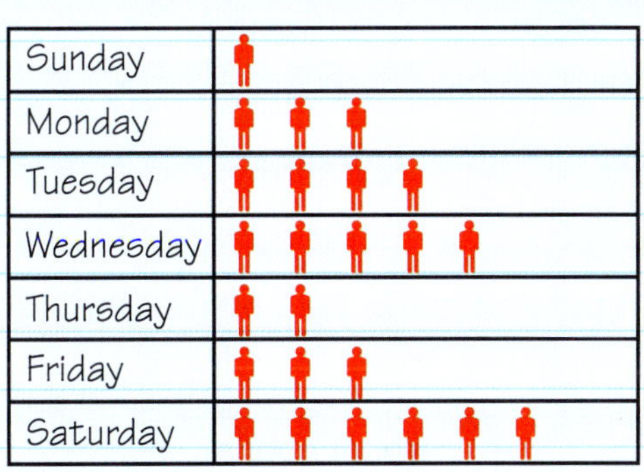

Represents 100 children

a. On which day did the smallest number of children go to the cinema?

b. How many children attended the cinema on Monday?

c. How many more children were there on Wednesday than on Friday?

d. How many children attended altogether on Friday and Saturday?

a. From the table it can be immediately seen that the smallest number of children attended on Sunday.

b. The pictograph uses a scale of one symbol to represent 100 children. There are three symbols for Monday. Hence the number of children present on Monday is 3 x 100 or 300 children.

c. There are two more symbols for Wednesday than Friday. This represents 2 x 100 children = 200 children.

d. There are nine symbols altogether for Friday and Saturday. Therefore there were 9 x 100 children = 900 children.

Exercise

1 If ☎ represents 20 calls:

a) How many calls does ☎ ☎ ☎ represent?

b) How many calls does  represent?

2  represents 60 ribbons sold.

a) How many ribbons sold does  represent?

b) Draw how you would represent 240 ribbons sold.

3 The data below shows the favourite fruits of a group of students.

banana	apple	cherry	grapes
banana	grapes	apple	banana
apple	cherry	banana	apple
grapes	banana	apple	cherry
cherry	apple	banana	apple

a) Draw a frequency table for the data.

b) Which fruit was the most popular?

c) Which fruit was the least popular?

4 The table below shows the number of books read by five children during the summer holidays.

Name	No. of books
Allison	12
Ronald	4
Grace	20
Greg	16
Barry	8

Using a rectangle to represent four books, show this information in a pictograph.

Key words
Bar graph
Axes

Unit 15.3 Bar graphs

In this unit you will:

- draw and use bar graphs.

It is not always convenient to use pictures to represent data. Another diagram which may be used is the **bar graph**. Bar graphs use scales to represent the information.

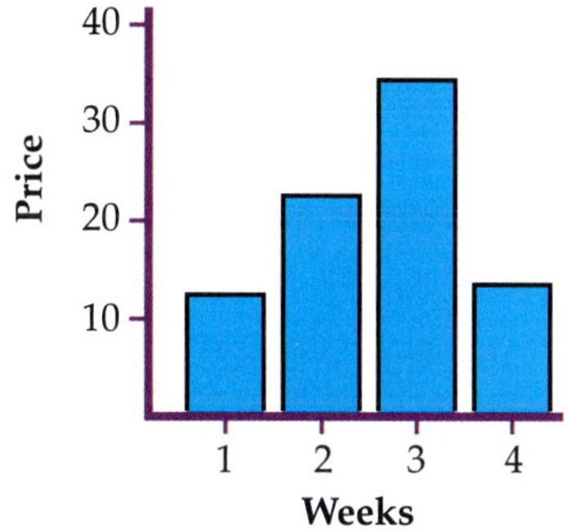

The bar graph uses bars of varying heights to show the information. The scales are shown against the **axes**.

Worked examples

The bar graph below shows the number of children in a secondary school.

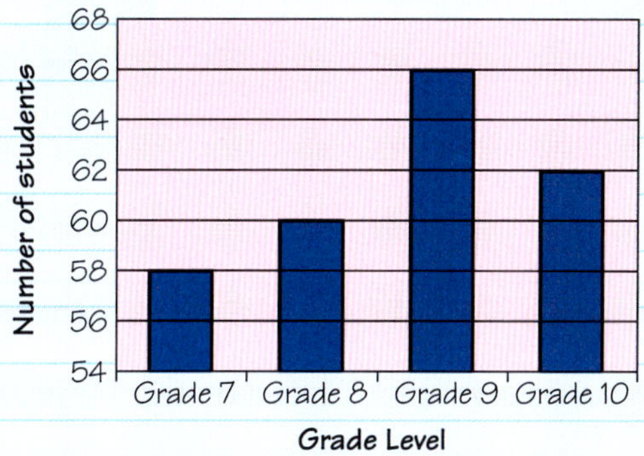

a. Use the bar graph to draw a frequency table.

b. How many children are there in the school altogether?

Discrete data, like this, can be counted and is usually recorded as a whole number. Continuous data (for example, height or speed) is measured, and so has an unlimited range of values.

a.

Form	Number of students
7	58
8	60
9	66
10	62

b. Total number of students: 58 + 60 + 66 + 62 = 246

Exercise

1. Fifty children were asked to state which sport they played on Saturday mornings.

   Draw a bar graph to show this information.

Sport	Number
Soccer	8
Hockey	15
Golf	6
Cricket	12
Badminton	9

2. The rainfall for the first six months of the year is shown in the table opposite.
   a) Draw a bar graph to show this information.
   b) Which month had the most rainfall?
   c) Which months had more rain than April?

Month	Rainfall (cm)
January	18
February	14
March	12
April	15
May	20
June	28

3. The bar graph opposite shows the number of points gained by five school sets for a spelling competition. The sets were Red, Green, Yellow, Blue and Purple.
   a) Which set won the competition?
   b) How many more points did Set Green receive than Set Blue?
   c) Which sets received a total of 80 points?

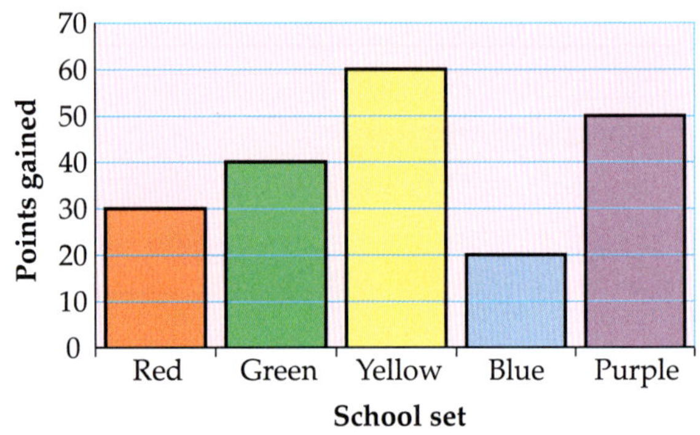

Project

You will need a magazine or newspaper which gives you the TV programmes for a week.

1. Work in small groups. Choose any day and allocate a television channel to each member.
   a) Draw up a table to show how much time each channel allocates to:
      - comedy
      - sport
      - drama
      - news and current affairs
      - quiz shows
      - educational programmes
      - children's programmes
      - other
   b) Discuss how you could show this information on one bar graph.
2. Draw up a questionnaire to find out the five most popular programmes amongst your classmates. Conduct a survey using your questionnaire. Tabulate your information and show it on a graph.
3. Survey your own TV-viewing time for a week in hours and minutes. Draw a bar graph for each of the members in the group. Discuss any patterns that emerge.

Key words
Pie chart
Angle
Sector

Unit 15.4 Pie charts

In this unit you will:

- draw and use pie charts.

The **pie chart** is another diagram that may be used to represent data. As its name suggests, it looks like a pie or pizza which has been cut into slices. The size (**angle**) of the slice tells us what fraction of the data is represented by each event. Each section of the pie chart is called a **sector**.

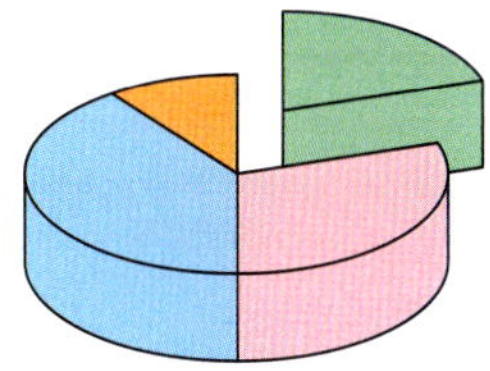

Worked examples

The pie chart shows how Mrs Brown spent the first day of her summer vacation.

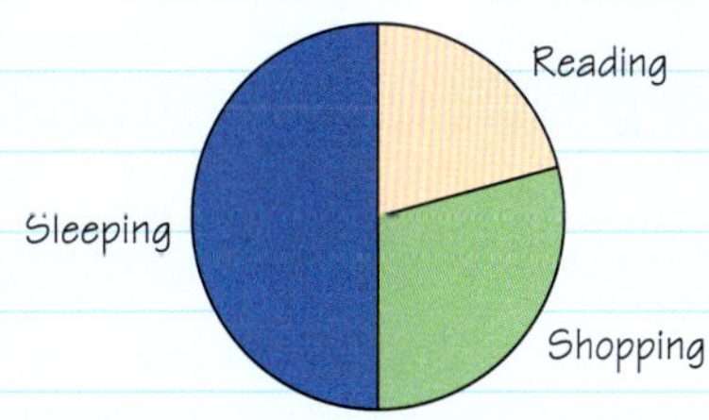

a. What fraction of the time did she spend sleeping?

b. If she spent five hours reading, how many hours did she spend shopping?

a. Since half of the pie chart represents sleeping, she spent half the day or 12 hours sleeping.

b. 12 hours sleeping plus 5 hours reading is a total of 17 hours. Hence, she spent 7 hours shopping.

Exercise

1 The favourite ice cream flavours of a group of children are shown in the pie chart below.

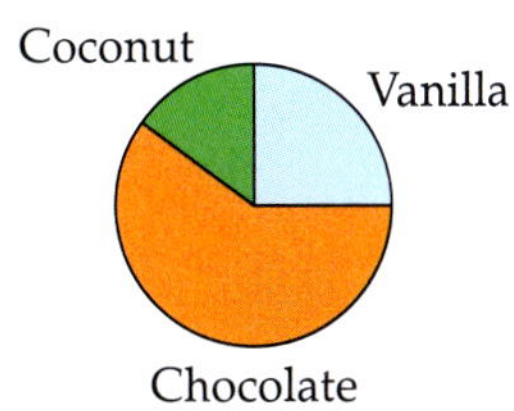

a) Which flavour was the most popular?

b) Which flavour was the least popular?

c) The sector representing vanilla is one-quarter of the circle. If there were 30 children who like vanilla, how many children were there in all?

2 The table below shows the countries represented in a group of 15 cricketers.

Country	Number
Barbados	4
Trinidad	3
Grenada	1
Guyana	2
Jamaica	5

a) Copy and complete the table to show the angles that will represent each country.

Country	Degrees
Barbados	
Trinidad	
Grenada	
Guyana	48°
Jamaica	

b) Draw a pie graph to show the information.

3 The table below shows the number of books read by a group of students last week.

Number of books	Frequency
One	15
Two	30
Three	10
Four	35

a) How many children were there in the survey?

b) How many degrees should be used to represent one child?

c) Draw a pie chart to show this information.

4 Jessica rolled a dice 16 times. She drew this pie chart to show her results:

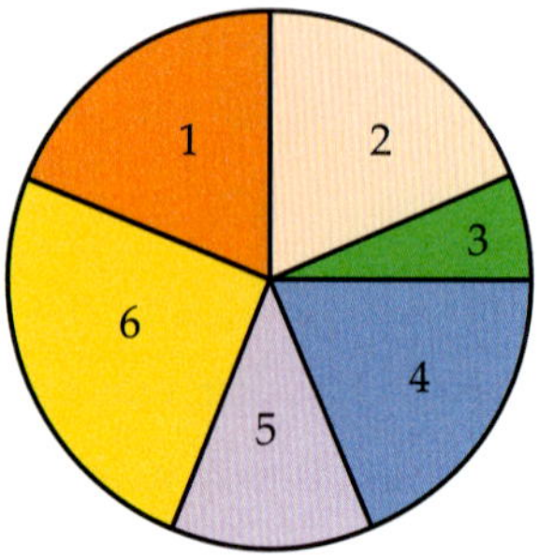

a) Which number did she roll most frequently?

b) Which number did she roll least frequently?

c) How many times did she roll a 6?

d) Do this experiment yourself. Draw a pie graph to show your results. What do you notice when you compare the pie graphs?

Case study - Dr Carlisle

Dr Carlisle is a medical doctor who specialises in the treatment of diabetes. The incidence of diabetes is particularly high amongst West Indians. If left untreated, diabetes may cause serious complications including high blood pressure, heart disease and blindness.

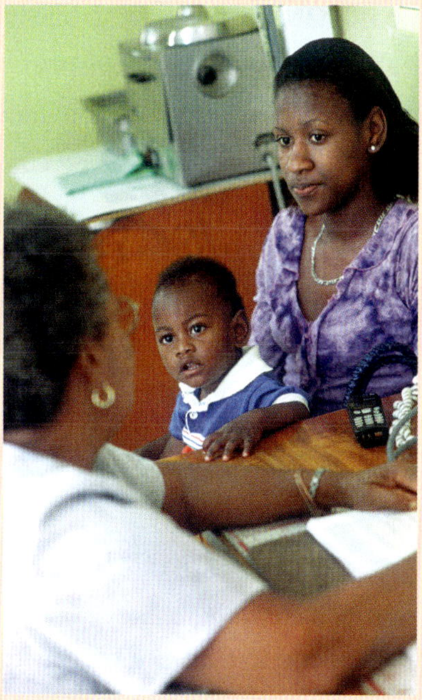

Part of Dr Carlisle's job is to try and reduce the number of cases of diabetes occurring each year. To assist with this, she collects data on patients by using a survey. She asks questions about their age, weight, height and medical history among other things. She then analyses the data and is able to make predictions to help in her work.

With this information, it is easy to tell the average age of persons with diabetes and what factors are likely to cause diabetes. For example, Dr Carlisle has discovered that the patients she sees for diabetes tend to be overweight for their age and height. She has also discovered that many people only start to get diabetes as adults. This information has helped her to educate patients to prevent the disease.

Dr Carlisle's statistics are collected from files and from patients themselves. When she works with the data, she enters it into her computer. She makes use of spreadsheets, statistical formulae and graph-generating software when she manipulates and analyses the data.

You can work like Dr Carlisle to compile some data. **As a class project**, collect data on the mass and height of five students. Create a table with these measurements. Check the measurements every month to see how they have changed. Draw a chart to compare some of these measurements.

Name of student		Month 1	Month 2	Month 3	Month 4	Month 5	Month 6
	Height						
	Mass						
	Height						
	Mass						
	Height						
	Mass						
	Height						
	Mass						
	Height						
	Mass						

Looking back

1 For each of the statements below, write down whether you would draw a pictograph, bar chart or pie chart to show the data collected.

a) Answers to the question 'What is your favourite pizza topping?'

b) The percentage of left-handed people in your class.

c) The population of the Caribbean islands.

d) The amount of money people spend on food each week.

2 Fifty people were asked what kind of take-away meal they preferred:

a) Draw a bar chart to represent this data.

b) What fraction of the group preferred pizza?

Type of take-away	Frequency
Chinese	12
Fish and chips	7
Hamburgers	11
Pizza	15
Hot dogs	5

3 A survey to find the favourite chip flavour of sixty children gave the results below.

Flavour	Cheese	Plain	Salt & Vinegar	Barbecue	Spicy
Frequency	20	8	10	15	7

Draw a bar graph to show this information.

4 The pie chart below shows how the children in a class travel to school. There are six children who travel by car and the size of the angle is 72°.

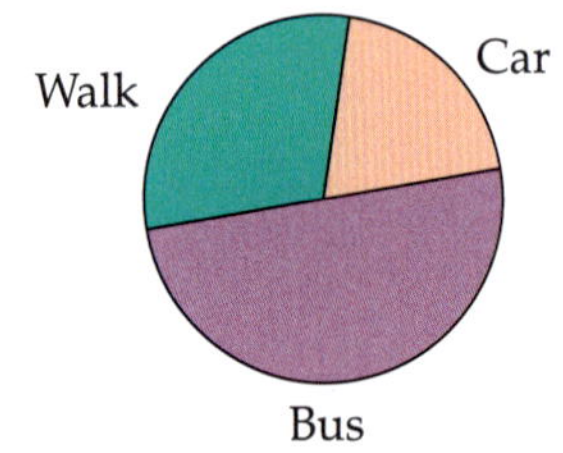

a) How many children are there in the class?

b) How many children either walk or travel by the bus?

c) If six more children travel by bus than walk, how many children walk?

5 The bar chart opposite shows the types of cars in a car park.

a) Which car was the most popular?

b) How many more Toyotas were there than Rovers?

c) How many cars were there altogether?

d) Using a symbol of ⊖ to represent 2 cars, draw a pictograph to show the information.

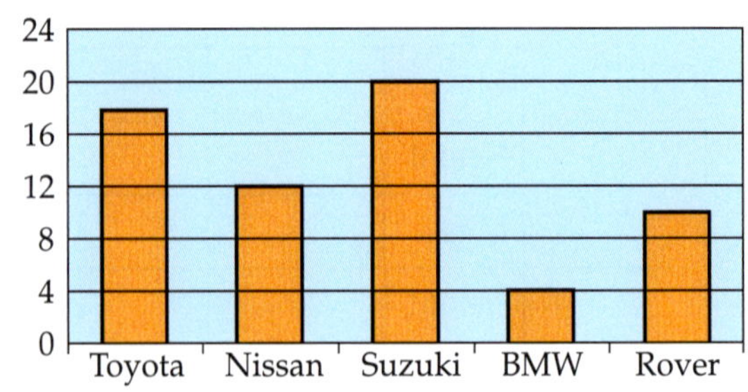

Project

1 a) On your way to school, record the colours of cars that you see. Copy the table below and complete it using tally marks to help you record the numbers. You may add other colours to the table.

Colour of car	Tally marks	Number
Red		
White		
Blue		
Silver		
Black		

b) What was the most popular colour seen?

c) Is this the most popular colour in your country?

2 The principal of the school wants to know the most popular drink sold at the school.

a) Explain how you will carry out this survey, by considering the following questions.

- Will you use a questionnaire, interview the students or observe what is being bought?
- How many days will you carry out the survey?
- How many students will you involve?

b) Conduct the survey and present your results to the class.

Chapter 16

Collecting and organising data

You have already seen how we can use different methods to collect data about a range of topics. You have also learned how to organise data and to use it to draw graphs. When data is represented on a graph, we can use the graph to identify patterns, or trends in the data. This chapter shows other methods that mathematicians and statisticians use to analyse data in order to calculate averages and identify trends. You will round off the chapter by planning, completing and reporting on your own statistical investigation.

Key words
Mode
Modal value

Unit 16.1 The mode

In this unit you will:

- determine the mode of a set of data.

After data has been collected and presented in a diagram, it must then be analysed to determine what information it really shows. In fact, most of the time, we don't have to collect data or draw diagrams as often as we need to interpret the information that has been collected by others. One trend that we can look for in data is the item or value that occurs the most often, that is, has the highest frequency. This is called the **mode** and the value is known as the **modal value**.

Worked example

The ages of a group of children are shown below.

10 8 6 8 9 8 7 10 6 8

Which age is the mode?

If we made a frequency table for the data we would have:

Age	6	7	8	9	10
Frequency	2	1	4	1	2

8 years has the highest frequency (4). This means that there are four children who are 8 years old and as this is the value that appears most often, 8 is the mode or modal age.

Note: When data is grouped and the ages are given as a class, for example 8–10, then the class with the greatest frequency is the modal class.

Exercise

1. Find the mode for each set of data:
   a) 4, 6, 3, 6, 8
   b) 2, 0, 4, 2, 0, 0, 7
   c) 3, 6, 9, 3, 5, 6, 3, 5
   d) 2.4, 3.6, 2.4, 5.3, 2.4, 4.2
   e) 4.5, 4.8, 4.6, 4.2, 4.6, 4.7

2. The colours of cars in a parking lot are shown below. What is the modal colour?

white	blue	green	white
green	white	red	blue
white	red	blue	green

Key words
Median
Range

Unit 16.2 The range and the median

In this unit you will:

- calculate the median for a set of data
- calculate the range for a set of data.

The **median** is another value that may be determined from data. The median is the middle value of a set of data which is ordered. The data may be arranged in ascending or descending order. If there is more than one value in the middle, we find the median by adding the two values and dividing by 2. The median may only be found for data which has a numerical value.

For a set of data, the **range** is the difference between the highest and lowest values. It tells us how spread out the data is.

Worked examples

1. The masses of five students are given in the table below. Determine:
   a. the student with the median mass
   b. the range of the masses.

Name	Jane	Bob	Jim	Sally	Ed
Mass (kg)	45	52	43	42	48

2. Determine the median of the set of numbers:
   7 4 20 6 14 11

1. a. First we must arrange the masses in numerical order.
   Starting with the smallest we have:
   42 43 45 48 52
   The mass in the middle is 45, which is Jane's mass.
   b. The largest mass is 52 kg and the smallest is 42 kg.
   Hence the range is 52 kg − 42 kg = 10 kg
2. First we have to order the numbers:
   4 6 7 11 14 20
   When there is an even number of values, there will be two middle numbers. To work out the median, we add them and divide by 2.
   $\frac{7+11}{2} = 9$

Exercise

1. Find the median and range for each of the following sets of numbers.
   a) 1, 1, 3, 3, 3, 4, 5
   b) 6, 6, 6, 7, 9, 9, 10, 10, 11
   c) 11, 11, 12, 13, 15, 16
   d) 0.4, 0.9, 1.1, 1.3, 1.5, 1.5
2. Toni asked her friends what mark they got for their last test. These marks are below.
   8 4 9 8 5 8 6 6 9 7 4
   a) What is the median mark?
   b) What is the range of marks?
3. The scores of eight cricketers were 65, 48, 77, 90, 53, 62, 9 and 35. What was the median score?

Unit 16.3 The mean

Key words
Mean
Average

In this unit you will:
- calculate the mean for a set of data
- compare the averages: mode, median and mean.

The **mean** is another calculation that may be used to analyse data which is numerical. The mean is found by using the formula:

$$\frac{\text{Total sum of all the data}}{\text{The number of items in the data}}$$

In everyday language, the mean is referred to as the **average**. But in fact, the mode, median and mean are all different types of averages.

Worked examples

1. The number of marbles bought by ten friends is shown below:

   6 7 5 9 3 8 2 9 6 5

   What was the mean number of marbles bought?
2. The mean mass of twenty mangoes was 0.35 kg. What was the total mass of the twenty mangoes?

1. The total number of marbles bought was:

   $6 + 7 + 5 + 9 + 3 + 8 + 2 + 9 + 6 + 5 = 60$

   Since there were 10 friends, the mean was $\frac{60}{10} = 6$.
2. The mean mass is 0.35 kg

   Therefore the total mass is $0.35\text{ kg} \times 20 = 7.0\text{ kg}$

Exercise

1. Find the mean of each of the following sets of numbers:

   a) 13, 8, 24, 19 b) 12, 6, 13, 5, 3, 9 c) 12, 8, 9, 11, 15
2. The costs of four caps were \$15.00, \$22.50, \$18.50 and \$16.00. What was the mean cost of a cap?
3. The mean of the numbers 11, 6 and x is 10. What is the value of x?
4. Ryan needs a mean of 55% in four tests to be promoted to a new class. His marks in three tests were 43%, 61% and 46%. What mark does he need in the last test to be promoted?
5. The six fastest times in seconds for Ato Bolden of Trinidad and Tobago to run the 100 metres are:

   9.86 9.87 9.88 9.89 9.90 9.92

   What is his mean time?

Key words
Primary source
Secondary source
Sample size

Unit 16.4 Planning an investigation

In this unit you will:

- plan a questionnaire, survey or experiment
- plan how to collect data including sample size.

You can use statistical methods to provide evidence to test a theory. First, say clearly what you want to investigate, then decide which data you need and how you will collect it. The diagram shows how an investigation might be carried out.

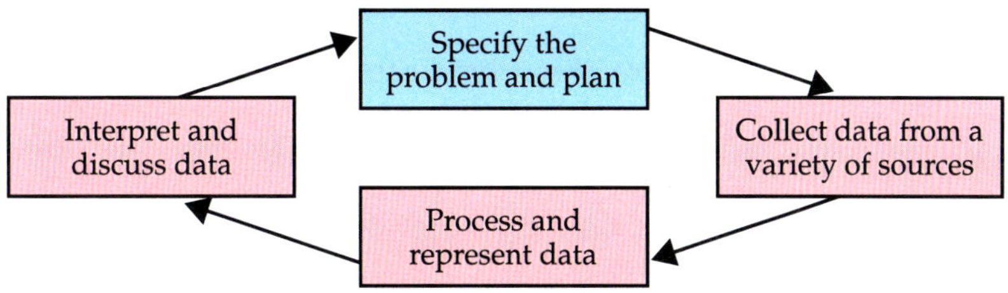

You can collect information from primary or secondary sources. A **primary source** is data you collect yourself. A **secondary source** is data that has already been collected. You can find secondary data from the Internet, a book or a CD-ROM, for example.

You will also need to decide on **sample size**, the number of people you will include. You need a sample size that is large enough to give reliable data, but not so large that it takes too long to collect and process the data.

Worked example

Nikki's friend says 'It's unfair – I can't do my homework, as I have no computer at home.' Nikki wants to find out if this problem affects a lot of pupils at her school.

a. What questions should she ask?
b. Where should she get her information?
c. Who should she ask?
d. How many pupils should she ask?

a. 1. Do you have a computer at home?
2. Do you have access to the Internet at home?
3. For how many hours per week can you use your computer?
4. How much of your homework needs the Internet?
5. Do you use computers anywhere else apart from school or home?

b. The school office might have data about the number of pupils with computers and Internet access at home. For questions 3–5, Nikki could carry out a survey.

c. She should ask pupils at her school from all year groups: both boys and girls, not just pupils in her class.

d. She should ask about 40–50 pupils.

This would be a secondary source.

This would be primary data.

Asking too few pupils will not give a variety of answers. Asking too many will take too long and make it difficult to make sense of all the information she has collected.

Exercise

1 Decide if the following would involve data from a **primary** or a **secondary** source:

a) The number of times a dice is rolled until a score of 30 or more is reached.

b) The relationship between the population of a country and the number of people who use the Internet.

c) The most popular author among pupils in your school.

d) The most popular author of children's books in the Caribbean.

e) The exam results of Form 4 pupils at your school.

f) The exam results of Form 4 pupils in the country.

g) The time it takes a cup of hot soup to reach room temperature.

2 Neil is carrying out a questionnaire about how often pupils play sport. He stands outside the school library at break, and asks six Form 1 girls and three Form 2 boys to fill in the questionnaire.

Name:		*Age:*	*Class:*	*Boy/Girl:*
How often do you play sport?	*Often* ☐	*Sometimes* ☐	*Never* ☐	*Depends* ☐
What sports do you play?	*Football* ☐	*Swimming* ☐	*Tennis* ☐	
Children don't play enough sports, do they?		*Yes* ☐	*No* ☐	

a) Show how his three questions can be improved.

b) Write down at least three mistakes Neil has made when carrying out his questionnaire.

c) Explain why Neil wants to collect information such as age, name and class.

3 'More mobile phones get stolen these days.'
Describe how you would collect data to find out if this statement is true.
Include detail about the sources of your data and sample size.

4 Plan an experiment to record the time it takes people to say the 5 times table, the 8 times table, and the alphabet backwards.
Include in your plan how you would find out if there is a relationship between the amount of time taken to say:

- the 5 times table and the 8 times table
- the 5 times table and the alphabet backwards
- the 8 times table and the alphabet backwards.

How accurate do the times need to be?

To draw frequency tables, your results will need to be grouped. Design class intervals for these frequency tables.

5 Lata thinks that more people will attend a football match when the team is in a higher football division. She intends to collect information about the number of people attending matches for teams that have moved divisions. Write a short plan to show Lata how to conduct her survey. Include details of where she will collect her information.

Investigation

Plan an investigation to test one of the following:

- Resting pulse rate is lower for older people.
- Expensive batteries last longer than cheap ones.
- CDs are more expensive at supermarkets than at town centre shops.

Key words
Data collection sheet
Sample size

Unit 16.5 Collecting data

In this unit you will:

- collect data using a suitable method.

In the previous unit we saw how to plan an investigation. In this unit, we will focus on methods of collecting data.

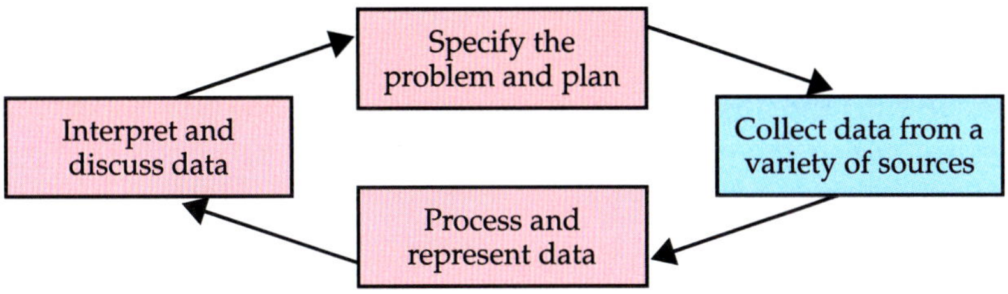

It is important to design a suitable **data collection sheet** that will allow you to record a range of data and process this data quickly. You will also need to decide how much data to collect – this is your **sample size**.

Worked example

A local business wants to find the best location to open a new sportswear shop.

a. Design a data collection sheet to find out:

i. ages and choice of shop for different age groups

ii. why people use certain shops

iii. how often they visit different shopping centres.

b. How many people should be questioned?

a. i.

Location \ Age	16–25	26–35	36–45	46–55	56–65	66+
Main street/local shops						
Out-of-town supermarket						
Internet						

Age categories should not overlap or have gaps.

ii. What affects your choice of shop? (Tick 1 box)

Convenience ☐ Cheap prices ☐ Transport ☐

Large choice ☐ Other ☐

Giving a small number of choices makes analysing the data easier. Include an 'other' box.

iii. How often do you buy sportswear?

More than once a month ☐

Once a month ☐

Less than once a month ☐

Categories should cover all possible answers.

b. Between 50 and 100 would be about right.

Asking a small number, 20 people, would not give enough data. Asking too many would take too long.

Exercise

1. 'Water boiled in the microwave cools down faster than water boiled in a kettle.' Plan an experiment to test this statement. Design a data collection sheet, and describe how you would collect the data.

2. Produce a plan to compare the performance of two football, cricket or rugby teams or two athletes. Which factors will you compare?
Design a data collection sheet.
How many years of data will you use?

3. Plan a survey to find the answer to this question: 'Does the time of day affect the amount or types of traffic on the road?'
Design data collection sheets to record:
   - the amount of traffic passing a point in a minute
   - the different types of vehicles
   - the number of passengers in a vehicle.

4. 'The frequencies of each letter used in the English language, from greatest to least are:
ETOAN IRSHD LCWUM FYGPB VKXQJZ'.
Design an experiment to test some or all of this statement.
Include a data collection sheet.
Where will you collect your data?
Will you use different sources?
How much data will you collect?

5. How long does it take to say the alphabet backwards?
Plan an investigation and design a data collection sheet.
How many times will you carry out this experiment?

6. Design a questionnaire to find out about the eating habits of pupils in your school.
Decide what type of information to collect and design a data collection sheet.
How many pupils will you ask?
Will you ask pupils of different gender and from different year groups?
Where will you carry out your survey?

7. Design an experiment to test if people can tell butter from different types of margarines and spreads. Design a data collection sheet.
How many people will you ask?
What different types of people will you ask?
Where will you carry out your survey?
What special equipment will you need?

Investigation

Carry out one or more of the above investigations, recording the information on your data collection sheet.

Key words
Continuous
Discrete

Unit 16.6 Displaying data

In this unit you will:

- draw bar charts and frequency diagrams for discrete and continuous data.

When you have collected the data you need, you can draw graphs and diagrams. It is important to choose appropriate diagrams for the type of data. Draw frequency diagrams for grouped **continuous** data, and bar charts for categorical and **discrete** data.

Worked example

Here are the results of a survey on Internet use:

For how long do you use the Internet each week?

Time (in hours)	0–2	2–4	4–6	6–8	8–10	10–12
Frequency	8	17	10	14	0	1

How many e-mails do you send each week?

Number of e-mails	0–4	5–9	10–14	15–19	20–24
Frequency	15	11	11	4	9

Where do you mainly use the Internet?

Location	School	Home	Library	Other
Frequency	16	31	2	1

Draw a suitable diagram for each table of results.

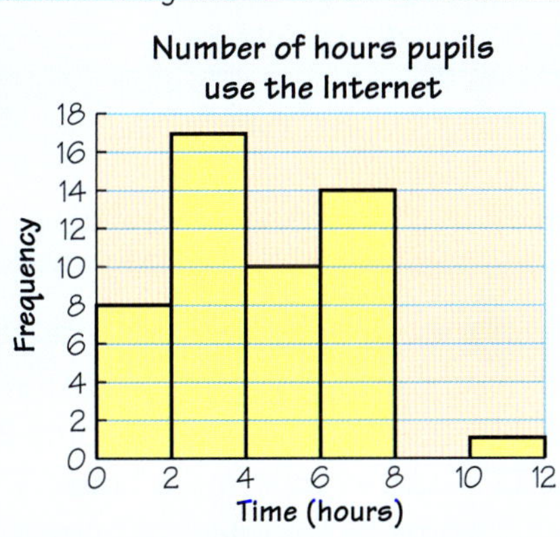

When the data is continuous, the bars touch, with the x-axis labels where the bars meet, not in the middle of the bar.

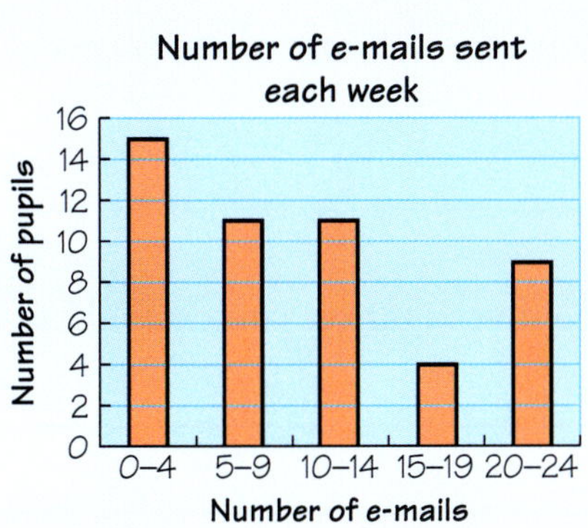

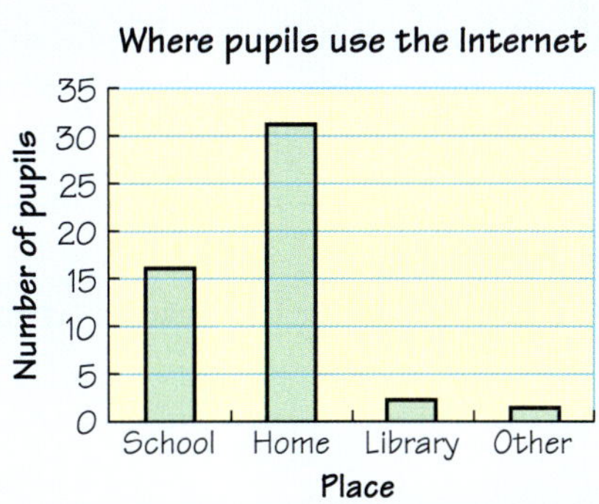

For grouped discrete data, and for categorical data, the bars do not touch.

Exercise

1. A teacher records the time spent by her pupils on an investigation. Draw a frequency diagram for her information.

Time (in hours)	0 – 1	1 – 2	2 – 3	3 – 4	4 – 5	5 – 6
Frequency	5	7	11	4	2	1

2. The school nurse measures the height of all the Form 1 boys. Draw a frequency diagram for this data.

Height (cm)	140 – 145	145 – 150	150 – 155	155 – 160	160 – 165	165 – 170	170 – 175
Frequency	1	5	17	27	32	16	2

3. Melissa conducts a survey about part-time work and pupils at her school. Draw a suitable diagram for each of her tables of results.

What type of work is your part-time job?

Type of work	Shop	Paper-round	Café/restaurant	Babysitting	Other
Frequency	14	6	9	7	3

How many hours do you work each week?

Number of hours	$0-1\frac{1}{2}$	$1\frac{1}{2}-3$	$3-4\frac{1}{2}$	$4\frac{1}{2}-6$
Frequency	15	11	12	1

How much do you get paid per hour?

Amount	\$0–\$4.99	\$5.00–\$9.99	\$10.00–\$14.99	\$15.00–\$19.99
Frequency	1	7	24	7

4. The times taken for 20 international swimmers in a long-distance swim are given below.

   18 hours 4 minutes, 14 hours 24 minutes, 12 hours 22 minutes, 13 hours 41 minutes, 11 hours 8 minutes, 9 hours 51 minutes, 13 hours 33 minutes, 12 hours 5 minutes, 13 hours 31 minutes, 10 hours 54 minutes, 17 hours 36 minutes, 16 hours 29 minutes, 11 hours 32 minutes, 13 hours 10 minutes, 12 hours 49 minutes, 8 hours 32 minutes, 11 hours 56 minutes, 16 hours 54 minutes, 11 hours 56 minutes, 14 hours 0 minutes

   Group the data into 2-hour class intervals, starting with 8 – 10 hours, then draw a frequency diagram for your table.

Investigation

Draw appropriate diagrams using the results from the investigation you carried out in unit 16.5.

Key words
Interpret
Cyclic

Unit 16.7 Communicating results

In this unit you will:

- communicate the results of a statistical enquiry
- interpret tables and diagrams for continuous data.

When you have calculated statistics and drawn diagrams to show your results, it is useful to **interpret** the results, and relate your findings to the original question.

Handling data is a **cyclic** process – this means that as you evaluate the results obtained, you may need to revisit the investigation with an improved plan, or you may find a related question to investigate.

Worked example

Shyam has also collected information about pupils using the Internet. Her investigation was 'Is there a difference between boys and girls in their use of the Internet?' Here is a table showing the statistics Shyam has calculated from her data.

	Boys	Girls
Mean (hours)	5	7
Median (hours)	4.5	6
Mode (hours)	5	2
Range (hours)	8	13

Shyam has also grouped her data and constructed frequency diagrams for boys and girls.

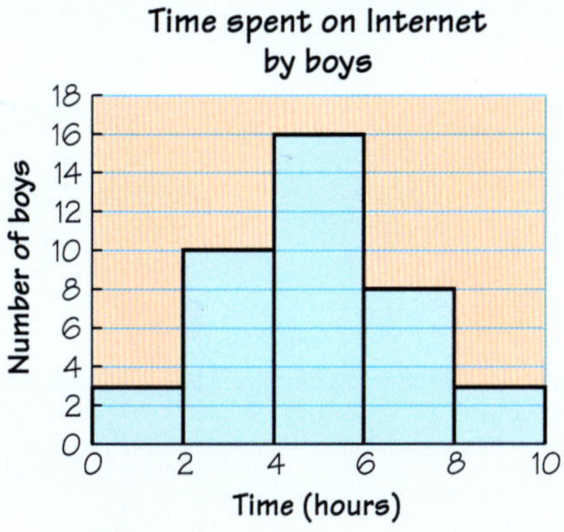

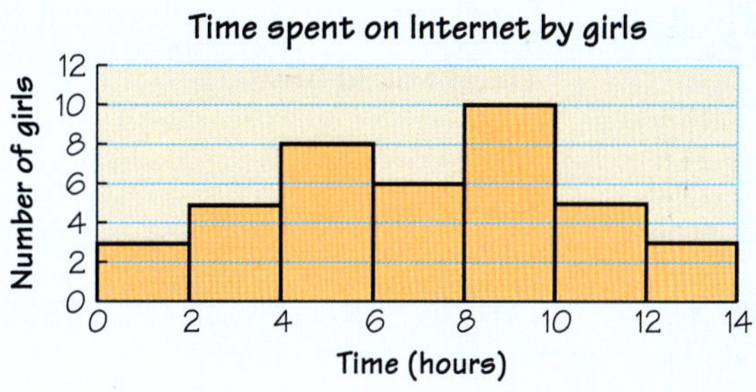

Use Shyam's results to answer her original question 'Is there a difference between boys and girls in their use of the Internet?'

The mean and the median show that girls use the Internet for longer lengths of time. The mode for the boys is higher than the mode for the girls, but the frequencies for the mode are probably low to draw any conclusions. The frequency diagrams show that the modal class for girls is 8 – 10 hours compared to 4 – 6 hours for boys.

The range of times for the girls is greater than the boys' range, showing a wider spread in the times they use the Internet for.

Remember the **mode** and the **modal class** are not the same.

Exercise

1 The table shows the usual time it takes people in part-time employment to travel to work.

Time (minutes)	Men (%)	Women (%)
1 – 11	32	41
11 – 20	22	26
21 – 30	10	11
31 – 40	2	2
40+	8	6

a) Which class interval shows the greatest difference between men's and women's travel times?
b) What percentage of women spend more than half an hour travelling to work?
c) Why do you think each set of percentages does not add up to 100%?
d) Write a sentence about the differences between the amount of time men and women spend travelling to work.

2 A group of pupils ran 200 metres.

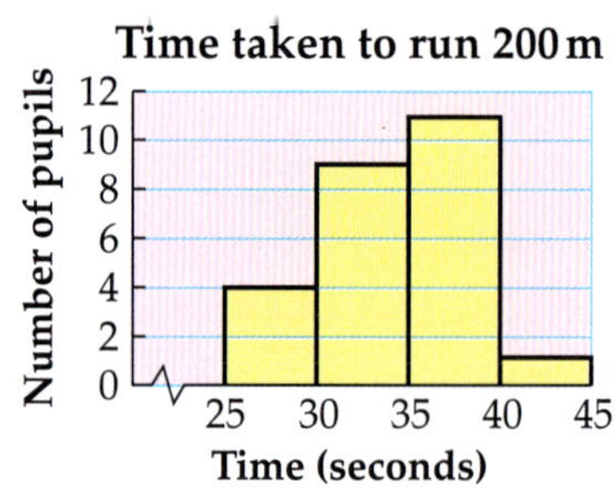

a) Write down the class intervals of this frequency diagram.
b) How many pupils ran the 200 metres?
c) What was the modal class?
d) How many pupils took more than 35 seconds to run 200 m?
e) How many pupils were in the interval 30 – 40 seconds?
f) Explain why the range may not be 45 − 25 = 20 seconds.
g) Jake ran 200 metres in 32 seconds. He says that 12 boys were faster than him. Explain why he may not be correct.

3 Ben would like to know 'Is there a difference in the heights of boys and girls in my year group?' He carries out an experiment to collect data. His results are below:

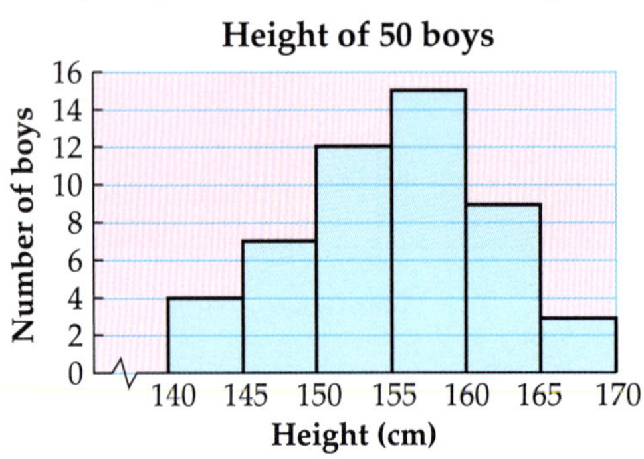

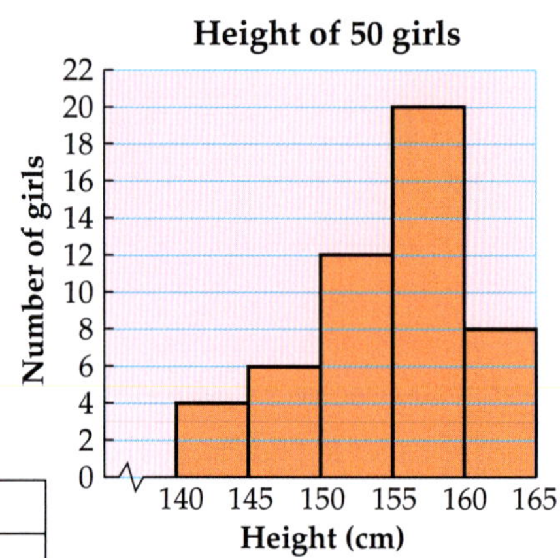

	Mean	Median	Range
Girls	154.5	155	18
Boys	155	155	28

Write a short report answering Ben's question.

Investigation

Use your results from your investigation in unit 16.6 to write a short report. Relate your report to your original question. Do your results suggest a further question you would like to investigate?

Looking back

1 A shoe store recorded the sizes of the shoes that it sold.

6	4	2	3	6	2	3	5
2	6	3	6	2	6	5	3

a) Find each of the three averages and the range for this data.

b) Should the store always stock the whole range of sizes? Explain your answer.

2 A brand of match states on the box that it contains 'on average' 50 matches per box. The contents of twenty boxes are counted as follows:

45	48	51	50	46	48	51	50	44	49
47	46	52	45	50	50	48	45	49	46

a) Make a frequency table for the data.

b) Find the mean, median and mode of the data.

c) Which of the three averages has been used to describe the 'average contents'?

d) Do you think this is a fair description of the number of matches in a box? Explain your answer.

3 a) Which of the three averages has a value that must appear in the data set?

b) Which of the three averages can have more than one value?

c) Which of the three averages may not be a number?

4 For each statement below, state which average should be used to give the result.

a) The average age of the boys in the school is 13 years 5 months.

b) The average number of hours a week that children watch television is 20.

c) The average rainfall for the month was 11.4 cm.

d) The average number of days in a month is 31.

e) The average number of cars per household is 2.

5 Faye's mathematics marks (as percentages) for ten tests are shown below.

16	25	32	32	66	68	70	72	75	76

a) Find the mode, mean and median for the set of marks.

b) Which of the averages – mode, mean or median – best describes her performance in mathematics?

6 The table below shows the total number of goals scored by teams in a hockey tournament.

Number of goals	4	5	6	7	8	9	10
Number of teams that scored this many goals	6	2	1	3	5	2	1

a) How many teams played in the tournament?

b) What was the modal score for the tournament?

c) What was the total number of goals scored?

d) What was the mean number of goals scored?

e) Jason has scored five goals or more in every hockey match he's played. Is he an above-average or below-average scorer? Why?

Project

Make a collection of graphs of all kinds from newspapers, magazines, books etc. (Try to include examples of all the different types you have studied.)

Discuss the graphs and decide which are the best ones. You could ask some of these questions to help you decide.

- Is the graph clear? What information is being shown? Is the graph accurate or does it mislead you?
- Is the scale being used clear and appropriate?
- Is the graph large enough to see all the important parts?
- Is the information shown important?
- Could I improve on this graph?

Try changing the presentation of some of the graphs. For example, could you change a bar chart to a pie chart?

- Is the new presentation better? If not, why not?
- If your graph is better – explain why you think so.

Project

Statistics provide the methods for collecting and then analysing information. The following suggestions could be followed up as individual projects or they could be group investigations.

Always try and ensure that the investigation shows evidence of the three stages of a statistical study:

- collecting data/information – by counting, tallying, measuring etc.
- displaying information – as tables, graphs, charts
- interpreting the information – looking for trends, seeing links between pieces of information, answering the question posed by the study.

Here are some ideas for surveys.

- What is the most popular ......? (drink, music, activity ...)
- What is the most common ......? (height, size of feet, car colour, name, letter, car number plate ...)
- Look at the traffic going past the school gates at different times of day. Does the percentage of cars and taxis change with the time of day?

Here are some statements to consider.

1 The proportion of the letter 'e' is characteristic of a language. The proportion of the letter 'e' in any passage of English is always the same, and is different from the proportion in, for example, French.

- Is this true or false?
- Could you tell what language something is written in, if you were told only the proportion of certain letters?
- Which letters are characteristic of a language in this way?

2 You can always tell how tall someone is by measuring their feet.

- How accurate is the link between foot size and height?
- Is it accurate enough for the statement to be true?

Test your knowledge

In all tests, choose the correct answer for multiple choice items. Working must be clearly shown for all other questions.

Test 1

1 Which of the following would be a suitable universal set for the subset {0, 3, 6, 9, 12}?

a) {even numbers} **b)** {odd numbers} **c)** {multiples of 6} **d)** {whole numbers}

2 Which of the following is true?

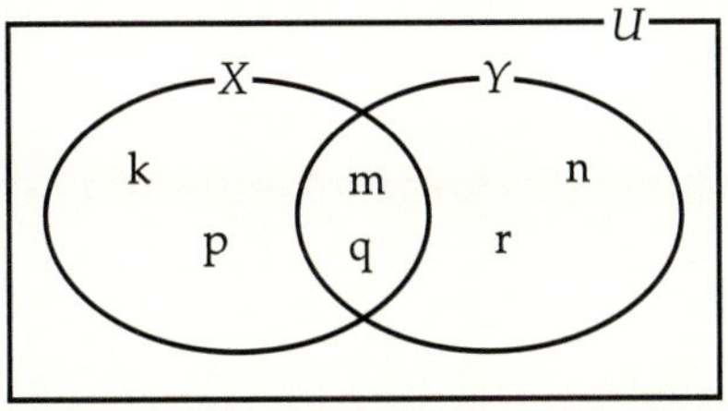

a) $X = \{k, p\}$, $Y = \{m, n, q, r\}$

b) $X = \{k, m, p, q\}$, $Y = \{m, n, q, r\}$

c) $X = \{k, m, p, q\}$, $Y = \{n, r\}$

d) $X = \{k, p\}$, $Y = \{n, r\}$

3 R = {students who enjoy Reggae music}
J = {students who enjoy Jazz music}
Which Venn diagram best describes 'Some students who enjoy Reggae also enjoy Jazz'?

a)
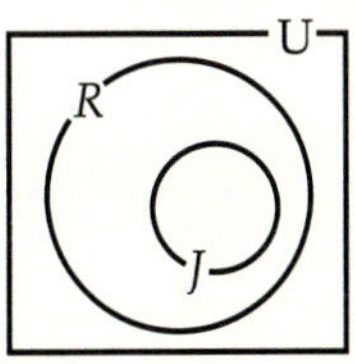

b)
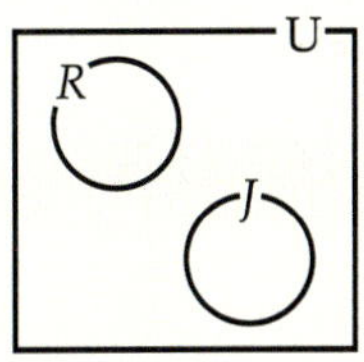

c)
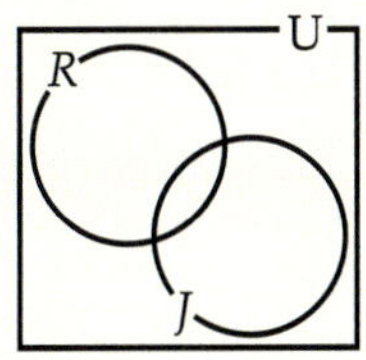

d)
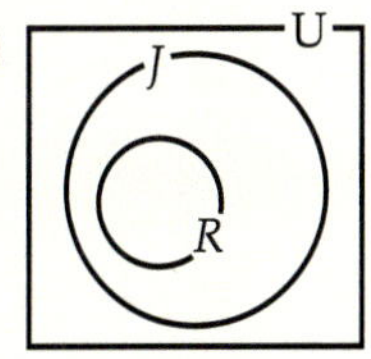

4 The Lowest Common Multiple of 6, 9 and 15 is

a) 3 **b)** 6 **c)** 90 **d)** 810

5 The Highest Common Factor of 20, 30 and 40 is

a) 120 **b)** 40 **c)** 20 **d)** 10

6 **a)** Make the largest number possible by writing the following pairs of digits in sequence: 13, 28, 51.

b) Give your answer correct to the nearest thousand.

7 **a)** Give a number which is greater than 15, less than 24 and is even.

b) Give a number which is odd, less than 100 but greater than 50, and a multiple of 7.

c) Give a number which is even, greater than 15 and a factor of 80.

d) Give two numbers which are prime, greater than 50 but less than 60.

8 Find **(i)** the HCF and **(ii)** the LCM of the following pairs of numbers:

a) 9 and 12 **b)** 16 and 40 **c)** 60 and 84

Test 2

1. A plot of land measures 37 m by 58 m. The area in m^2 is

   a) 5 080 b) 2 146 c) 1 946 d) 176

2. Penny buys gifts costing \$2.75, \$2.09 and \$1.25. The change from \$10.00 will be

   a) \$6.09 b) \$4.91 c) \$3.91 d) \$3.09

3. Which of the following gives the best percentage mark?

   a) 30 out of 60 b) 36 out of 72 c) 44 out of 90 d) 21 out of 40

4. Calculate:

   a) $4\,962 + 307 + 2\,185 + 792$ b) $3\,008 - 859$
   c) 416×38 d) $1\,752 \div 73$

5. Jane is in training for a sponsored walk. On Monday she walks 3 km, on Wednesday 4 km, and on Friday she walks 6 km.

   a) How far has she walked altogether?
   b) How far must she walk on Sunday to make a total of 20 km for the four days?

6. Find:

   a) 25% of \$10.00 b) 50% of \$120.00 c) 66% of \$300.00

7. Find the value of each of the following:

   a) 50% of 55 b) 32% of 46 c) 16% of 300

8. Bill saved \$5.45 by buying a tennis racket at a sale for \$45.90. What was the price before the sale?

   a) \$40.45 b) \$50.35 c) \$50.45 d) \$51.35

9. Calculate:

   a) 4 × \$1 + 4 × 25c + 2 × 10\$ + 1 × 5c
   b) 3 × \$1 + 2 × 25c + 5 × 10\$ + 5 × 5c
   c) 2 × \$1 + 6 × 25c + 10 × 10\$ + 3 × 5c
   d) 1 × \$1 + 10 × 25c + 10 × 10\$ + 10 × 5c

10. How much is left from \$50 after buying 3 books at \$12.65 each?

    a) \$12.05 b) \$13.15 c) \$37.35 d) \$38.65

Test 3

1. After a discount of \$15 is given, the net price of an article is \$32.50. What was the price before the discount?

   a) \$17.50 **b)** \$32.35 **c)** \$32.65 **d)** \$47.50

2. What is the net price of an article marked at \$65 if a discount of 10% is given?

   a) \$58.50 **b)** \$71.50 **c)** \$6.50 **d)** \$56.50

3. A dress bought for \$60 is sold at a profit of 20%. What is the selling price?

   a) \$40 **b)** \$48 **c)** \$72 **d)** \$80

4. Mrs Jones' helper is paid \$800 per month with board and lodging. If board and lodging is valued at \$95 per week, what will it cost to keep her helper for 1 year?

5. Mr Chin is a salesman. His salary is \$1 500 per month plus 3% of the value of his sales. Calculate his salary in a month in which sales are \$135 000.

6. **a)** A discount of 5% is given on car mirrors. How much will a buyer pay for a car mirror marked at \$25?

   b) A car dealer makes a profit of 25% on car tyres. What is the selling price of a tyre which he buys for \$205?

7. How many days are there in $3y$ weeks?

   a) $10y$ **b)** $21y$ **c)** $37y$ **d)** $73y$

8. $(7x + 2y) + (3x + 4y) =$

   a) $16xy$ **b)** $10x + 6y$ **c)** $16 + xy$ **d)** $9xy + 7xy$

9. If r is an even number the next even number above r is

   a) $2r$ **b)** $r + 1$ **c)** $r + 2$ **d)** $\frac{r}{2}$

10. $2 \times 3 \times y \times 2 \times y \times y =$

    a) $12y^3$ **b)** $36y$ **c)** $7y^3$ **d)** $6y^4$

11. $3 \times 7a - 4 \times 2b =$

    a) $21a - 8b$ **b)** $13ab$ **c)** $9a - 2b$ **d)** $17a - 2b$

12. The solution of $2x - 6 = 10$ is $x =$

    a) 2 **b)** 3 **c)** 5 **d)** 8

Test 4

1. a) Hooper scored y runs. Lara scored 4 more runs than Hooper. How many runs did Lara score?

   b) In a one-day cricket match, Pakistan scored 15 runs less than West Indies. If West Indies scored x runs, how many runs did Pakistan score?

   c) The width of a rectangular room is a metres. The length is 3 times the width. What is the length?

   d) The Maths class contains 5 more girls than boys. There are d girls in the class. How many boys are there in the class?

   e) John's score in an examination was p marks. If this score is decreased by q, what is the final score?

2. Form equations and find solutions for the following:

   a) Mary's age is x years. Mary's age increased by 12 is the age of Jane. If Jane is 22 years old, how old is Mary?

   b) Henry is thinking of a number, x. Two less than three times the number is 37. What is the number?

3. The formula $s = c\,(1 + r)$ may be used to find the selling price, s, of a printer where c = cost price, r = rate of profit. Find the selling price of a printer, if the cost is \$505 and the rate of profit is 15%.

4. 78 mm = 7.8 cm, 144 m = 14.40 km, 12 m = 12 000 cm. How many of these statements are true?

   a) 0   b) 1   c) 2   d) 3

5. Which statement is true? 0.74 g =

   a) 0.0074 mg   b) 7.4 mg   c) 740 mg   d) 7 400 mg

6. If the length of a box-kite is 0.6 m and its width is 0.5 m, the area is:

   a) 3.0 m^2 b) 0.30 m^2 c) 30 cm^2 d) 300 cm^2

7. Theo's house is 11 minutes from the bus station. She is 8 minutes early for a 10.15 a.m. bus. At what time did she leave her house?

   a) 9.56 a.m.   b) 10.04 a.m.   c) 10.07 a.m.   d) 10.34 a.m.

8. Find the area of a rectangle 40 cm wide and 4 m long (answer in square metres).

9. I take 3 parcels to the post office. One has a weight of 2 kg 763 g, the second 3 kg 87 g and the third 4 kg 462 g. Find the total weight in kg.

Test 5

1. How many prime numbers are there between 27 and 36?

   a) 1 b) 2 c) 4 d) 8

2. Which of the following numbers must be divisible by both 2 and 5?

   a) 23 752 b) 43 825 c) 25 347 d) 26 870

3. Mrs Hall was given a 10% discount on a dress whose price was $70. How much did she pay?

   a) $60 b) $63 c) $77 d) $80

4. If * means 'multiply the first number by 4 then subtract the second', the value of 6 * 5 is

   a) 1 b) 4 c) 19 d) 26

5. Mrs Walkes changed $20 to EC dollars at a rate of $1.00 = EC $4.60. The amount of money she received was

   a) $15.40 b) $20.46 c) $24.60 d) $92.00

6. The difference between 10% of 50 and 50% of 10 is

   a) 0 b) 1 c) 5 d) 10

7. My car needs new tyres every 16 000 km, new plugs every 10 000 km and an oil change every 5 000 km. How many kilometres will my car have travelled before all three changes are needed at the same time again, if these were all changed at the same time on the last occasion?

8. If P = {M, A, T, H} and Q = {M, A, T, H, S}, which of the following statements are true?

   a) $P \subset Q$ b) $Q \subset P$ c) $P = Q$ d) $P \in Q$ e) $Q \in P$

9. How much change does mother receive from $100 after buying 12 lead pencils, 12 coloured pencils, 2 white shirts, 2 pairs of khaki shorts and one 3 pcs lunch kit?

FOR SCHOOL

Khaki, Blue and White Shirts	*Now 2 for* **$15.00**
Khaki Shorts	*Now 2 for* **$17.00**
Lead Pencils	*12 for* **$2.00**
Coloured Pencils	*12 for* **$2.95**
3 pcs LUNCH KITS *Reg $19.95*	*Now* **$11.95**
Children's Raincoats	**$14.95** *or 2 for* **$25.00**

10. If a 10-second television advertisement costs $350, how much will a 16-second advertisement cost? (Assume the same cost per second.)

Test 6

1. How many obtuse angles are there in the following set?

   {60°, 140°, $1\frac{1}{2}$ right angles, 3 right angles, 200°}

   a) 0 **b)** 1 **c)** 2 **d)** 3

2. At 5 p.m. the size of the smaller angle between the hands of the clock is

   a) 30° **b)** 90° **c)** 135° **d)** 150°

3. Given the set of angles {reflex, right, obtuse}, which of the following lists the angles in the correct order, smallest first?

   a) {reflex, obtuse, right}

   b) {right, obtuse, reflex}

   c) {obtuse, right, reflex}

   d) {right, reflex, obtuse}

4. Draw and measure:

   a) an acute angle

   b) an obtuse angle

   c) a reflex angle

5. Through how many degrees does the minute hand of a clock turn when it moves from:

   a) 6 to 9

   b) 10 to 7

   c) 1 to 12?

6. Draw one net for each of the following shapes.

   a) cube

   b) cone

   c) cuboid

7. How many faces do the following have?

   a) a cuboid

   b) a square-based pyramid

   c) a triangular-based pyramid

Test 7

1. '10 less than y' may be written as

   a) $10 - y$ b) $y - 10$ c) $\frac{10}{y}$ d) $\frac{y}{10}$

2. If $\frac{n}{2} = 12$, then $n =$

   a) 6 b) 10 c) 14 d) 24

3. If $a = 3$, $b = 0$ and $c = 4$, what is the value of ab^3c^2?

   a) 0 b) 24 c) 48 d) 304

4. If a number 100 is first decreased by 10% and the resulting number is increased by 10%, the new number is:

   a) 99 b) 100 c) 110 d) 120

5. You are facing north and turn clockwise through a right angle. You are then facing

   a) east b) south c) west d) east or west

6. An article bought for \$150 is sold at a loss of 10%. What is the selling price?

   a) \$135 b) \$140 c) \$160 d) \$165

7. This shape is called a

   a) parallelogram b) trapezium
   c) pentagon d) rectangle

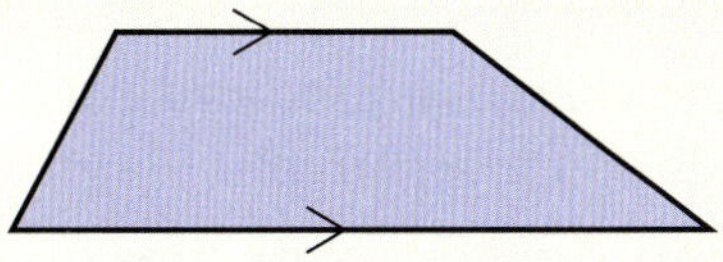

8. Measure the angle $X\hat{Y}Z$.
   Its size is about

   a) 35° b) 65° c) 90° d) 100°

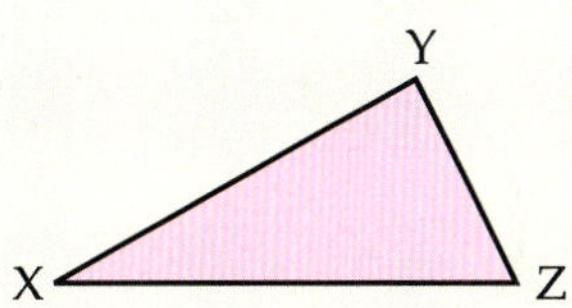

9. The 'square metre' is used to measure

   a) weight b) solid volume c) length d) area

10. a) Simplify $3b + 2c + 6c + 4b - b - 2c$.

    b) If $x = 4$, $y = 5$ and $c = 12$, evaluate $x^2 + 2y - \frac{c}{3}$

11. John deposits \$$d$ in the bank each week. How much will he have in the bank after 20 weeks? If he withdraws \$30, how much will he have left in the bank? If the amount left is equal to \$150, how much did he deposit each week?

12. Calculate 't' in each of the following:

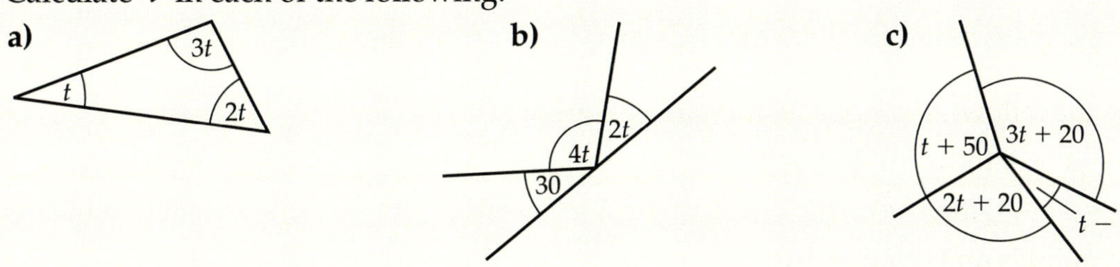

Test 8

1 $2 \rightarrow 6,\ 4 \rightarrow 12,\ 8 \rightarrow 24,\ 16 \rightarrow 48.$
The rule for the above mapping is
a) multiply by 2 **b)** multiply by 3 **c)** add 4 **d)** add 6

2 $y = 4x + 1$. If $x = 2, y =$
a) 2 **b)** 9 **c)** 12 **d)** 43

3 Which of the following points is on the y-axis?
a) (0, 2) **b)** (2, 0) **c)** (2, 2) **d)** (2, 3)

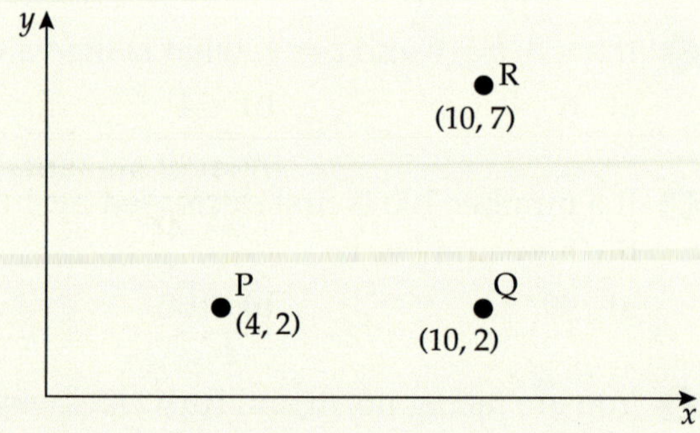

Use the diagram to answer questions 4 and 5.

4 The length of PQ is
a) 4 **b)** 6 **c)** 10 **d)** 14

5 The length of RQ is
a) 3 **b)** 5 **c)** 8 **d)** 9

6 A car travels d km in t hours. If $d = 40t$, what is d when $t = 2$?
a) 20 **b)** 38 **c)** 42 **d)** 80

7 A clock repairer charges a fixed price of \$5 for each clock he repairs, plus \$2.50 for every hour he spends on the repair job.
Draw a graph to show this, with time up to 12 hours on the horizontal axis and cost on the vertical axis.
a) How much does a repair that takes $7\frac{1}{2}$ hours cost?
b) If a repair cost \$16, how long, to the nearest half hour, did it take?

8 Copy and complete this mapping diagram.

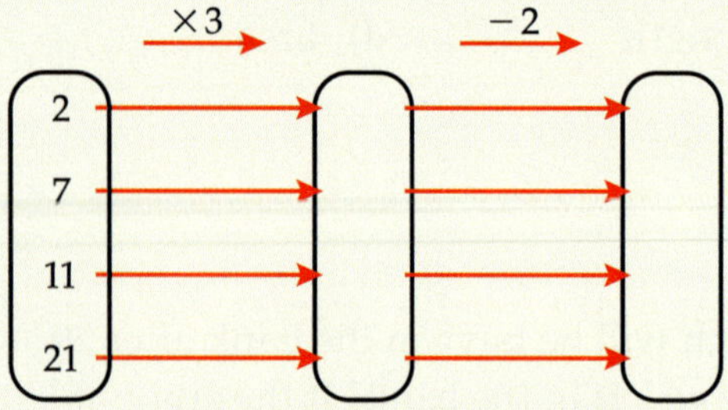

9 **a)** Plot the points P(4, 3), Q(2, 3), R(3, 5).
b) Draw the triangle PQR. What kind of triangle is it?
c) How many units are there in the perpendicular from R to PQ?

10 Draw the following graphs on the same diagram.
a) $y = x + 3$, for $x = 0, 1, 2, 3, 4$
b) $y = 2x - 1$, for $x = 1, 3, 5$.
What point is on both graphs?

Test 9

1. In 6 innings a batsman had a mean score of 30. What was his total score in the 6 innings?

   a) 24 **b)** 30 **c)** 36 **d)** 180

2. In 5 tests Joanne had scores of 55, 64, 60, 60 and 85. The range of scores was

   a) 30 **b)** 55 **c)** 60 **d)** 85

3. 180 students were asked to state their favourite subject. The numbers choosing each subject are shown in the table.

Subject	Number of students
Maths	44
English	30
Science	36
History	28
Geography	42

   Draw a pie chart to show this information.

4. The ages of 36 students in Forms 2–4 are:

15	15	16	15	16	15	14	14	16
14	14	14	14	14	13	15	16	14
14	15	14	13	15	14	16	13	14
14	15	16	14	15	14	13	15	13

   a) Arrange the data in a frequency table showing age, tally and frequency.

   b) Draw a bar graph for the data.

5. The table below shows the rainfall during a year in Trinidad and Tobago.

Month	Jan	Feb	Mar	Apr	May	Jun	Jul	Aug	Sep	Oct	Nov	Dec
Rain (in cm)	25	24	15	27	28	30	30	84	55	60	70	35

   Calculate the mean rainfall for that year.

6. The mean height of three girls is 165 cm. The heights of two of them are 160 cm and 158 cm. How tall is the third girl?

Test 10

1. If the product of two whole numbers is even, then:
   a) both must be even
   b) one number must be even
   c) both numbers must be odd
   d) one number must be odd

2. A complete turn is divided into 15 equal angles. The size of each angle is:
   a) 6° **b)** 12° **c)** 15° **d)** 24°

3. U is the set of {2, 3, 4, _ _ _ _ 10} and two of its subsets are P {odd numbers} and Q {prime numbers}. Which of the following is true?
   a) $P \subset Q$ **b)** $Q \subset P$ **c)** $P = Q$ **d)** $P \sim Q$

4. The solution of $2p - 7 = 9$ is:
   a) $p = 1$ **b)** $p = 2$ **c)** $p = 8$ **d)** $p = 16$

5. $7\frac{1}{2}\%$ of \$140 is
   a) \$5.25 **b)** \$10.50 **c)** \$21 **d)** \$105

6. The perimeter of a rectangle 4 cm long and 8 mm broad is
   a) 12 mm **b)** 24 mm **c)** 48 mm **d)** 96 mm

7. 'When 3 is multiplied by t the result is $\frac{1}{2}$' can be written as
   a) $3 + t = \frac{1}{2}$ **b)** $\frac{1}{2}t = 3$ **c)** $3t = \frac{1}{2}$ **d)** $3 \times \frac{1}{2} = t$

8. Maya's test scores were: Maths 35 out of 60; English 46 out of 80; History 31 out of 50; Science 40 out of 70. In which subject did she get the best percentage?
   a) Maths **b)** English **c)** History **d)** Science

9. **a)** On graph paper, plot the points P, Q and R with coordinates (−2, 0), (0, 2) and (2, 0).
   b) Plot the point S, so that PQRS is a square.

10. Jill drew a rectangle 9 cm long and 4 cm wide. Harry drew a square of equal area. What was the length of side of Harry's square?

Answers

Chapter 1

Page 1

1 Students' verbal answers
2 Students' answers

Page 3

1 **a)** 2 hundreds
b) 6 thousands, 1 unit
c) 1 ten thousand
d) 2 hundred thousand, 1 ten
e) 80 or 8 tens
f) 1 hundred
g) 1 million, 8 hundreds
h) 7 hundred thousands, 9 thousands

2 **a)** St John
b) River Island
c) St Peterstown

3 **a)** 158 000, 247 000, 268 000, 1 310 000, 2 700 000
b) Jamaica – two million, seven hundred thousand
Trinidad and Tobago – one million, three hundred and ten thousand
Barbados – two hundred and sixty-eight thousand
St Lucia – one hundred and fifty-eight thousand
Belize – two hundred and forty-seven thousand

Page 4

1

Number	Whole	Natural	Odd	Even
8	✓	✓		✓
17	✓	✓	✓	
45	✓	✓	✓	
257	✓	✓	✓	
21	✓	✓	✓	
29	✓	✓	✓	
$\frac{7}{8}$				
$4\frac{1}{5}$				
64	✓	✓		✓
1	✓	✓	✓	

Number	Prime	Rectangular	Composite
8		✓	✓
17	✓		
45		✓	✓
257	✓	✓	
21		✓	✓
29	✓		
$\frac{7}{8}$			
$4\frac{1}{5}$			
64		✓	✓
1			

2 No, it is divisible by one number – itself
3 Unlimited; so large it can't be quantified
4 No; 2 is even and prime
5 **a)** 2, 3, 5, 7, 11, 13, 17, 19
b) 8, 9, 10, 12, 14, 15, 16, 18, 20, 21, 22, 24, 25, 26, 27, 28
c) −21, −23, −25, …, −37, −39
d) 18, 20, 21, 22, 24, 25, 26, 27, 28, 30, 32, 33, 34, 35, 36, 38, 39
6 **a)** even **b)** odd **c)** even
d) even **e)** even **f)** odd

Page 7

1 **a)** 4 **b)** 9 **c)** 27
d) 49 **e)** 125 **f)** 64
g) 1 000 **h)** 243 **i)** 1 000 000 000
2 **a)** 10^2 **b)** 10^4 **c)** 10^6 **d)** 10^5
e) 10^9 **f)** 10^1 **g)** 10^0
3

Index form	Expanded form	Square numbers
1^2	1×1	1
2^2	2×2	4
3^2	3×3	9
4^2	4×4	16
5^2	5×5	25
6^2	6×6	36
7^2	7×7	49
8^2	8×8	64
9^2	9×9	81
10^2	10×10	100

4

Index form	Expanded form	Cube numbers
1^3	$1 \times 1 \times 1$	1
2^3	$2 \times 2 \times 2$	8
3^3	$3 \times 3 \times 3$	27
4^3	$4 \times 4 \times 4$	64
5^3	$5 \times 5 \times 5$	125
6^3	$6 \times 6 \times 6$	216
7^3	$7 \times 7 \times 7$	343
8^3	$8 \times 8 \times 8$	512
9^3	$9 \times 9 \times 9$	729
10^3	$10 \times 10 \times 10$	1 000

5 **a)** 8 **b)** 1 000
c) 4 **d)** 1 025

Page 9

1 **a)** 5 000 **b)** 23 000
c) 3 600 **d)** 9 700
e) 200 **f)** 3
g) 70 000 **h)** 900
i) 300 **j)** 55
k) 5 000 **l)** 647 000
m) 8 000 **n)** 33

2 **a)** 470 000 **b)** 905 000
c) 700 000 **d)** 342 000
e) 4 950 **f)** 3 200
g) 7 010 000 **h)** 350 000

3 **a)** 6.76 **b)** 0.009 17
c) 0.455 **d)** 8
e) 5 **f)** 0.007 25
g) 6 **h)** 4
i) 9 **j)** 0.5
k) 0.033 **l)** 0.007

4 132.58 m

5 8.9 cm, 25.4 cm

6 1.408 kg

Looking back (page 10)

1 **a)** 2 370 **b)** 10 027
c) 9 372 000 **d)** 173 333

2 **a)** 600 **b)** 60
c) 600 000 **d)** 6

3 The face value is what the number is on its own (3, 8, etc.). The place value is how many units, tens, hundreds etc., the number is worth. And the value of a digit is the same as the place value.

4 **a)** 9, 54, 63, 99 **b)** 2
c) 2, 17, 23, 61, 83, 107, 127

5 Using a computer

6 **a)** 2^4 **b)** 4^3 **c)** 10^5
d) 12^6 **e)** 5^2 **f)** 10^3

7 **a)** 2 **b)** 8 **c)** 1 024
d) 10 000 **e)** 1 **f)** 10
g) 125 **h)** 27 **i)** 729

8

Kilometres	Metres	Centimetres	Millimetres
12	12 000	1 200 000	12 000 000
99	99 000	9 900 000	99 000 000
100	100 000	10 000 000	100 000 000
2 200	2 200 000	220 000 000	2 200 000 000

Chapter 2

Page 12

1 **a)** 56 734 **b)** 1 300 341 **c)** 333 330
d) 19 148 **e)** 110 127

2 **a)** 382 **b)** 2 074 **c)** 23 076
d) 6 088 **e)** 19 435

4 **a)** Up and Away
b) Fly Anytime
c) 489 437
d) **i)** 64 642 **ii)** 19 284

5 **a)** \$84 373 **b)** \$651 900
c) \$351 449, \$351 451

6 **a)** eight thousand one hundred, eight thousand five hundred, one thousand three hundred and ten, three thousand nine hundred and fifty, two thousand seven hundred, eleven thousand two hundred and twenty-two
b) 35 782
c) **i)** 9 912 **ii)** 1 250

Page 13

1 **a)** 59 736 **b)** 15 288 **c)** 85 275
d) 4368 **e)** 14 061 **f)** 123 400

2

	Dividend	Divisor	Quotient	Remainder
a)	3 465	15	231	0
b)	835	26	32	3
c)	4 000	32	125	0
d)	11 430	18	635	0
e)	6 031	48	125	31
f)	900	100	9	0

4 **a)** 224 **b)** 26
c) 12 375 **d)** 1R26

5 **a)** 895 **b)** 2 320 **c)** 26
d) 71 **e)** Students' own answers

6 **a)** 7 litres **b)** 70 litres **c)** 38.5 litres

7 **a)** 1 856 kg **b)** 9.5 kg

8 **a)** 192 **b)** 48 000 m **c)** 148 800 m

Page 16

1 **a)** 18 **b)** 24 **c)** 120 **d)** 5
e) 0 **f)** 10 **g)** 11 **h)** 5

2 **a)** 30 **b)** 48 **c)** 42 **d)** 120
e) 2 **f)** 4 **g)** 3.2 **h)** 3

3 **a)** 31 **b)** 52 **c)** 11
d) 75 **e)** 11 **f)** 118
4 **a)** 27 **b)** 25 **c)** 6 **d)** 7
e) 19 **f)** 18 **g)** 5 **h)** 15
i) 19 **j)** 4 **k)** 44 **l)** 20
m) 8 **n)** 40 Extra answer: 9
5 **a), b), f), g)** and **h)** are true.

Page 17

1 **a)** $13 300, $9 000, $19 500, $12 500
b) $13 300, $9 000, $19 500, $12 500
c) $13 000, $9 000, $19 000, $13 000
2 **a)** 90, 84 **b)** 10, 13
c) 1 200, 1 573 **d)** 2 100, 1 836
e) 64.5, 56 **f)** 37, 35
g) 6, 3, 373, 264, 8.5, 2
3 **a)** 10 km **b)** 20 km **c)** 40 km
4 **a)** 21 000 **b)** 560 000
c) 100 **d)** 10
5 If you estimate the answer, you will notice if you add extra digits or miss a digit on the calculator, ending up with a number that is much too large or too small.

Page 19

1 **a)** 13, 14, 15 **b)** 56, 48, 40
c) 16, 32, 64 **d)** 60, 2, 62
e) 8, 13, 21 **f)** 0, 1, 0
2 **a)** 31, 38 **b)** 32, 64
c) 20, 25 **d)** 22, 18
e) 16, 4 **f)** 15, 23
g) 108, 216 **h)** 11, 19
i) 100, 125, 150 **j)** 13, 21, 34
3 **a)** 9 **b)** 12 **c)** 27
4 **b)** 16 **c)** 21
d)

Number of beach-huts	1	2	3	4	5	6	7
Number of matchsticks	6	11	16	21	26	31	36

5
1
1 1
1 2 1
1 3 3 1
1 4 6 4 1
1 5 10 10 5 1
1 6 15 20 15 6 1

6 **a)** 9
b) There would be months with no hours of daylight.

Page 21

Number	Factors	Prime factors product
12	1, 2, 3, 4, 6, 12	$2^2 \times 3$
18	1, 2, 3, 6, 9, 18	2×3^2
27	1, 3, 9, 27	3^3
32	1, 2, 4, 8, 16, 32	2^5
45	1, 3, 5, 9, 15, 45	$3^2 \times 5$
48	1, 2, 3, 4, 6, 8, 12, 16, 24, 48	$2^4 \times 3$
65	1, 5, 13, 65	5×13
55	1, 5, 11, 55	5×11
72	1, 2, 3, 4, 6, 8, 9, 12, 18, 24, 36, 72	$2^3 \times 3^2$

Page 22

1 **a)** 7 **b)** 2 **c)** 6
d) 3 **e)** 16 **f)** 1
2 3 cm
3 23

Page 23

1 **a)** 54 **b)** 120 **c)** 132
d) 294 **e)** 900 **f)** 100
2 Friday next week
3 180

Page 24

1 **a) i)** $\frac{1}{2}$ **ii)** $\frac{3}{4}$ **iii)** $\frac{2}{5}$ **iv)** $\frac{1}{5}$
b) i) $\frac{1}{2}$ **ii)** $\frac{1}{4}$ **iii)** $\frac{3}{5}$ **iv)** $\frac{3}{5}$
c) i) 1 : 1 **ii)** 3 : 1 **iii)** 2 : 3 **iv)** 1 : 3
d) i) 1 : 1 **ii)** 1 : 3 **iii)** 3 : 2 **iv)** 3 : 1
2 **a)** 2 red, 6 yellow **b)** 4 red, 8 yellow
c) 16 red, 4 yellow **d)** 14 red, 6 yellow
3 **a)** 3 : 1 **b)** 1 : 5 **c)** 5 : 2 **d)** 3 : 2
e) 2 : 3 **f)** 1 : 10 **g)** 1 : 9 **h)** 1 : 1
i) 15 : 7 **j)** 5 : 9
4 Students' own answers
5 3 : 4
6 1 : 4
7 **a)** 1 : 3 **b)** 7 : 13 **c)** 11 : 9

Page 26

1 **a)** $8, $16 **b)** $9, $15 **c)** $18, $6
d) $14, $10 **e)** $3, $21 **f)** $19, $5
2 **a)** $30, $6 **b)** $16, $20 **c)** $9, $27
d) $33, $3 **e)** $10, $26 **f)** $18, $18
3 14
4 $18.67, $37.33, $56.00
5 **a) i)** 24 **ii)** 32 **iii)** 16 **iv)** 40
b) i) 8 **ii)** 12 **iii)** 6 **iv)** 13
6 **a)** 14 **b)** 63 **c)** 49
7 Harry 24; Tom 12
8 Bruce $20; Eric $25; Babsie $35

Page 29

1 Any combination of 2, 5, 1 in left column and 3, 6, 7, 4 in right column OR
3, 1, 4 in left column and 6, 2, 5, 7 in right column

3 **a)** 8 **b)** 5 **c)** 27

Page 30

1 **a)** 22 **b)** 102
c) Number of triangles + 2

2 **a)** 8, 7; minus 1, plus 3
b) 13, 21; add the previous two numbers
c) 125, 216; cube numbers
d) 14, 17; add 3

3 **a)** 15 **b)** 201
c) Multiply by 2, then add 1

Looking back (page 31)

1 **a)** 123 **b)** 20 803 **c)** St Patrick

2 **a)** 115 **b)** 12 **c)** 5

3 No. $5 - 3 \neq 3 - 5$; $12 \div 4 \neq 4 \div 12$

4 **a)** 37, 46 **b)** 16, 22 **c)** 125, 625

5 **a)** 30 **b)** 2 **c)** 25
d) 132 **e)** 18

6 **a)** 6 : 7 **b)** 5 : 7 **c)** 2 : 9
d) 5 : 2 **e)** 2 : 5 **f)** 2 : 5

7 **a) i)** 6 **ii)** 27
b) i) 30 **ii)** 70

8 92

9 2250

10 1st, 3rd and 4th are women, 2nd and 5th are men.

Chapter 3

Page 32

1 **a)-c)** Students' own answers.

2 **a)** the first 5 letters of the alphabet **b)** 4 pets
c) 5 islands in the Caribbean
d) the first 4 months of the year
e) even numbers greater than 10

3

Column A	Column B
{5,6,7}	{3,2}
{2,3}	{14,12,11,9,10}
{cat, dog, pig, bird}	{6}
{10,11,12, 13,14}	{0,1,2,3,4,5,6}
{whole numbers from 1 to 6}	
{a}	{c,a,t}
{ }	{2,4,6,8}
{3,5,7,9,11,13,15}	{whole numbers from 20 to 25}

Sets matched with blue lines are equivalent

4 Students' own answers.

5 **a)** {2, 3, 5, 7}
b) {Monday, Tuesday, Wednesday, Thursday, Friday, Saturday, Sunday}
c) {St Lucia, Grenada, Dominica, St Vincent}
d) {6, 8, 10, 12, 14}
e) {12, 15, 18, 21, 24}

6 **a)** $\notin$ **b)** $\in$ **c)** $\in$
d) $\notin$ **e)** $\in$

7 **a)** infinite **b)** finite
c) infinite **d)** infinite
e) finite **f)** infinite
g) finite **h)** finite
i) finite **j)** finite

8 **a), c), d)**

Page 34

1 When the elements of A are contained in M and the number of elements of A is less than the number of elements in M

2 **a)** { }, {a}, {b}, {c}, {a ,b}, {a, c}, {b, c}, {a, b, c}
b) all except the last

3 **a)** $\subset$ **b)** $\not\subset$ **c)** $\not\subset$
d) $\subset$ **e)** $\subset$

4 **a)** {}
b) {}, {table}
c) {}, {pencil}, {pen}, {eraser}, {pencil, pen}, {pencil, eraser}, {pen, eraser}, (pencil, pen, eraser}
d) {}, {2}, {3}, {4}, {5}, {2, 3},{2, 4}, {2, 5}, {3, 4}, {3, 5}, {4, 5}, {2, 3, 4}, {2, 3, 5}, {3, 4, 5},{2, 4, 5}, {2, 3, 4, 5}
e) {}, {penny}, {dollar}, {penny, dollar}

5 Students' own answers.

6 Students' own answers.

7 **a)** {3, 5, 7, 11}
b) {3, 9}
c) { }

8 **a)** even numbers
b) numbers divisible by 11
c) capitals of Caribbean islands
d) continents

Page 36

1 Students' own answers.

2 **a)**

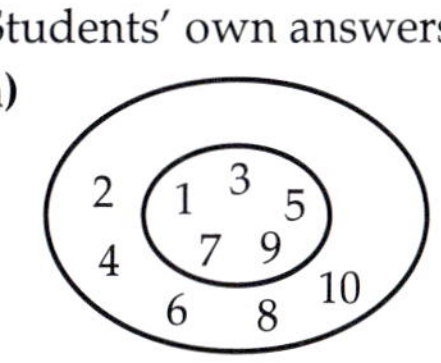

b)

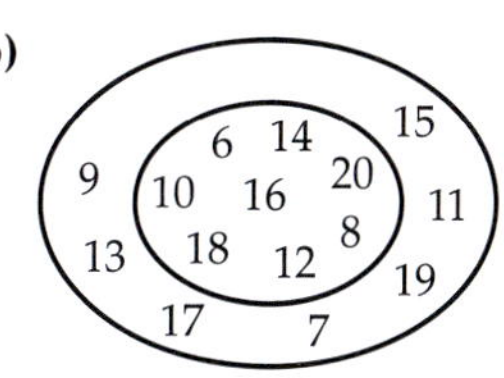

c)

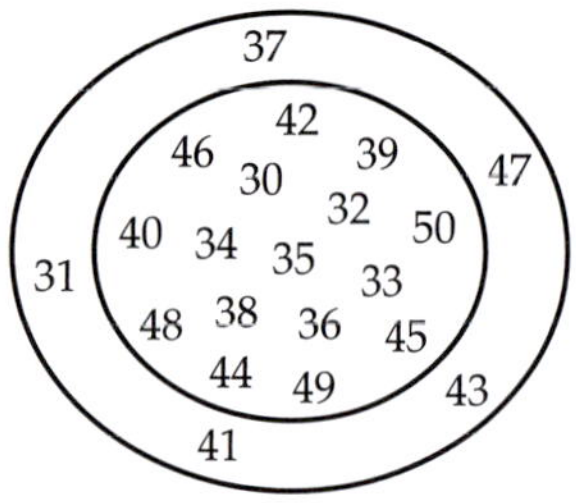

d)

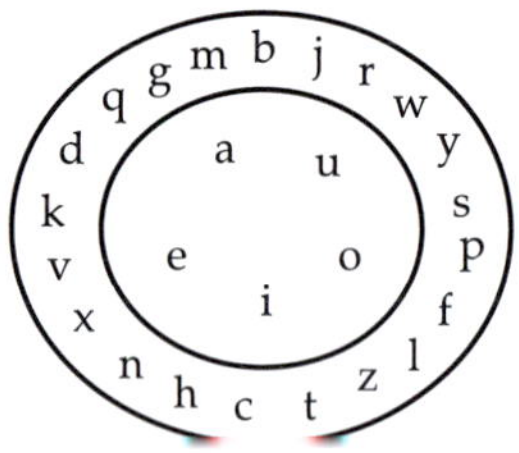

e)

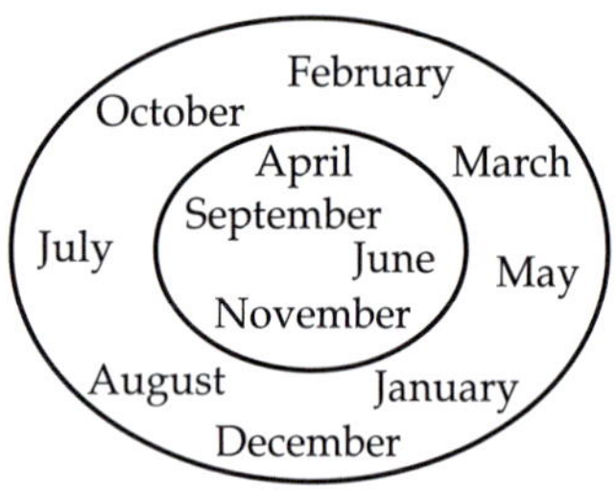

f)

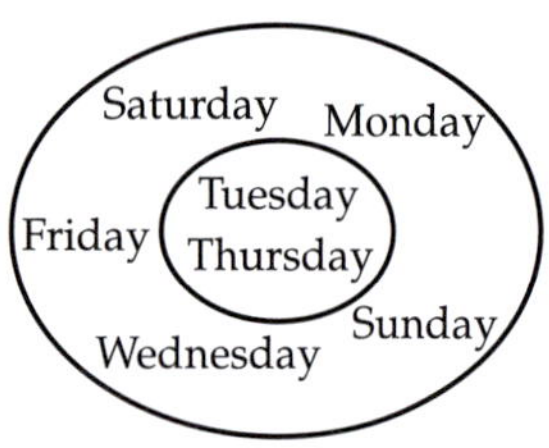

g)

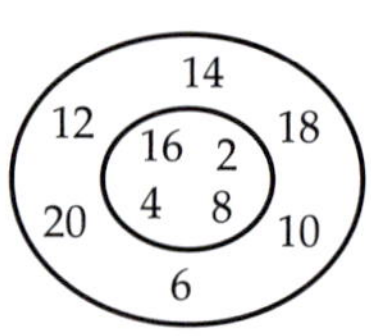

h)

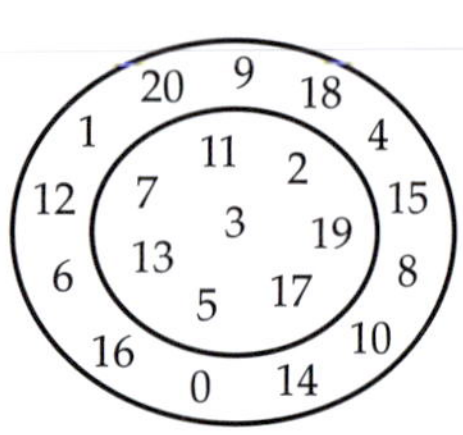

i) Students' own answers.

j) Students' own answers.

3 a)

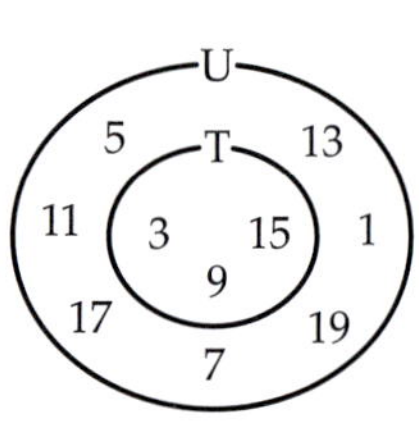

b)

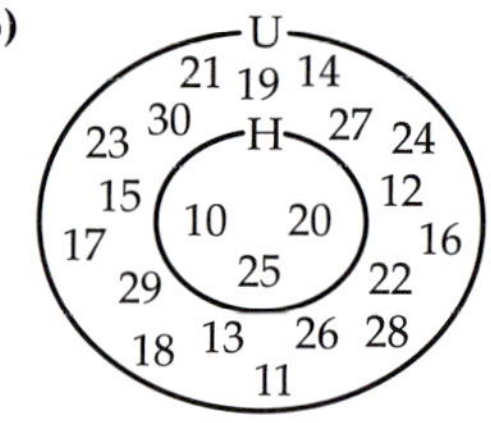

4 **a)** 50 **b)** 35 **c)** 15 **d)** 11

5 **a)**

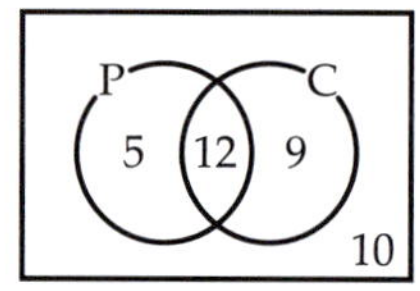

b)

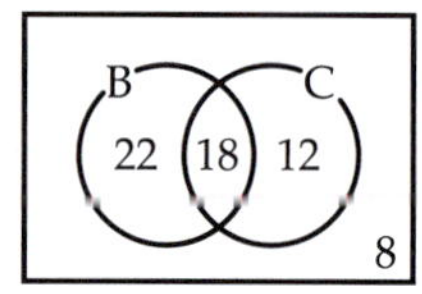

Looking back (page 38)

1 **a)** {C, A, R, I, B, E, N} **b)** {1, 2, 3, 4, 6, 9}
c) Students' own answers. **d)** {}
e) {20, 21, 22, 24, 25, 26, 27, 28}

2 Students' own answers.

3 **a), b)**

4 **a)** {}, {1}, {2}, {3}, {4}, {1, 2}, {1, 3}, {1, 4}, {2, 3}, {2, 4}, {3, 4}, {1, 2, 3}, {1, 3, 4}, {2, 3, 4}, {1, 2, 4}, {1, 2, 3, 4}.
b) 16 subsets

5 Inside M – January, March, May, July, August, October, December
Outside M – February, April, June, September, November

6 **a)** Inside small circle – 4, 8, 12, 16, 20, 24, 28
Outside small circle – 2, 6, 10, 14, 18, 22, 26, 30
b) Inside small circle – E, X, T, R, A, O, D, I, N, L, Y
Outside small circle – B, C, F, G, H, J, K, M, P, Q, S, U, V, W, Y, Z

Revise and consolidate 1

Page 39

1 **a)** four thousand seven hundred and thirty-two
b) eight hundred thousand and seventy-three
c) three thousand and nine
d) one million, three hundred and one thousand and fifty-two

2 **a)** three tens or 30 **b)** three units or 3
c) three thousand or 3000
d) three hundred thousand or 300 000

3 **a)** 26 **b)** 24 **c)** 27 **d)** 37
e) 0 **f)** 8 **g)** 28 **h)** 6
i) 11 **j)** 32 **k)** −2 **l)** 29

4 **a)** + − **b)** − + −
c) − + + **d)** ÷ ×
e) × ÷ **f)** ÷ ÷ ×

5 a) 17
b) e.g. 27
c) any prime number e.g. 2
d) 2
e) 1
f) e.g. 9

6 a) 4 b) 6 c) 7 d) 4

7 a) 144 b) 468 c) 240 d) 156

8 390

9 74

10 a) 5^5 b) 3^3 c) 12^3 d) 1^7
e) 7^2 f) 3^6 g) 25^5 h) 10^6
i) 1000^2 j) 2^4

11 a) $2 \times 2 \times 2 = 8$
b) $4 \times 4 = 16$
c) $5 \times 5 \times 5 = 125$
d) $10^0 = 1$
e) $8 \times 8 \times 8 \times 8 \times 8 \times 8 = 262\,144$
f) $10 \times 10 \times 10 \times 10 \times 10 = 100\,000$
g) $9 \times 9 \times 9 \times 9 \times 9 \times 9 = 531\,441$
h) $15 \times 15 \times 15 \times 15 \times 15 = 759\,375$

12 a) {a, b, c, d, e} b) {1, 3, 5, 7, ...}
c) $C \subset B$ d) { }
e) $c \notin \{a, b\}$ f) $B \simeq X$

13 a)

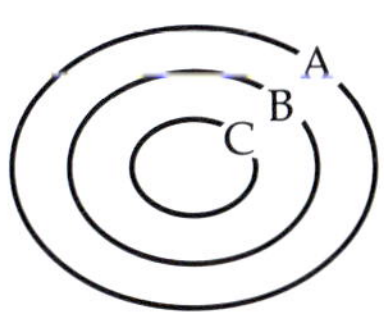

A = scholars on island
B = scholars in town
C = scholars in school

b)

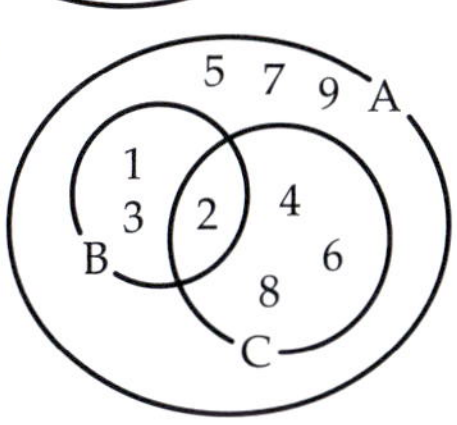

A = {1, 2 ...... 9}
B = {1, 2, 3}
C = {2, 4, 6, 8}

14 A = {4, 5, 6}
B = all natural numbers less than 7

15 a) no b) yes
c)

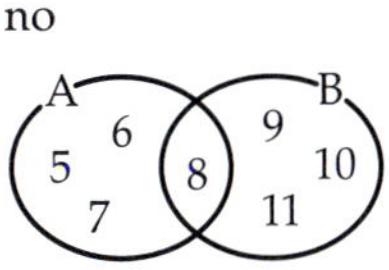

d) X = {8}

Chapter 4

Page 42

1 a) Furlong – the distance a team of oxen could plough without resting.
b) The span is the width of a man's hand with the fingers stretched out as far as possible.
c) The mina is a unit of weight that was about one pound but has varied to as much as 2 pounds.
d) Horsepower is the power exerted by a horse in pulling 550 pounds at a rate of 1 foot per second.

2 Pupils' own answers

3 a) All the measurements will be different for each person.
b) A standard unit is needed where all the measurements will be the same.
c) 12 inches or 30.48 cm

4 Mass was measured by comparing two objects to see which was heavier. Later a set of stones was kept for weighing. Different stones were kept for weighing different things. Some civilisations used a wheat seed as the smallest unit of weight.

5 Students' own answers

Page 44

1 a) centimetres b) grams
c) kilometres d) kilograms
e) metres f) millimetres

2 a) 5 m b) 120 mm c) 3 g

3 a) Students' own estimates
b) Students' own estimates
c) Students' own estimates
d) Students' own estimates
e) Students' own estimates
f) Students' own estimates

Page 45

1 Students' own answers
2 Students' own answers
3 The litre bottle holds more.
4 Students' own answers
5 Trudi's sister is taller.
6 Approximately \$200 (\$182.82)
7 Approximately 10 kg (9.08 kg).

Page 47

1 a) 5 °C b) 15 °C c) 7 °C
d) −5 °C e) −4 °C f) 0 °C

2 a) 10 °C b) 22 °C c) 36 °C

3 a) 73.4 °F b) 113 °F c) 140 °F

4 −50 °F, −12 °C, 6 °C, 50 °F, 12 °C, 29 °C, 100 °F, 90 °C, 212 °F

5 a) 30 °C b) 86 °F

Page 48

1 a) 08 : 25 b) 22 : 45
c) 02 : 18 d) 03 : 05
e) 12 : 30 f) 24 : 00 or 00 : 00

2 a) 9:35 a.m. b) 4:20 p.m.
c) 8:10 p.m. d) 6:55 a.m.
e) 11:00 p.m. f) 12:00 midnight

3 a) 16 hours 5 mins b) 32 hours 15 mins
c) 32 hours 15 mins d) 32 hours 30 mins

Looking back (page 49)

1 **a)** 15 feet 5 inches **b)** 4 m 70 cm
2 Approximately 60 inches
3 3.4 kg
4 3:13 p.m. or 15:13
5 2:30 p.m.
6 2:45 p.m. or 14:45
7 6 °C
8 **a)** February **b)** 17 °C **c)** 8 °C
9 **a)** 570 mm **b)** 98 mm **c)** 0.61 m
d) 12 000 m **e)** 0.84 km

Chapter 5

Page 51

1 **a)** 2.3 cm **b)** 0.96 cm
c) 34 cm **d)** 1.2 cm
e) 5 500 cm **f)** 820 cm
g) 900 000 cm **h)** 45 000 cm
2 **a)** 23.8 m **b)** 0.056 m
c) 6.5 m **d)** 0.08 m
e) 75 m **f)** 5 400 m
g) 9 000 m **h)** 330 m
2 **a)** 80, 86, 92 **b)** 2, 0, $^{-}2$
3 **a)** 4.5 km **b)** 0.08 km
c) 0.00048 km **d)** 0.0065 km
e) 0.72 km **f)** 2.5 km
g) 0.00053 km **h)** 0.000004 km
4 **a)** cm **b)** km **c)** mm or cm
d) cm **e)** m
5 Students' own estimates

Page 53

1 **a)** 21.0 cm **b)** 19.3 cm
c) 50 cm **d)** 12.6 cm
e) 51 mm **f)** 30.2 cm
2 **a)** 280 mm **b)** 46 cm
c) 27 cm **d)** 32.5 cm
3 62.5 m
4 Students' own answers
5 25 m
6 20 cm
7 80 cm

Looking back (page 54)

1 5.71 m
2 562 m
3 1.44 m
4 30 pieces
5 68 mm
6 150 m
7 8 km
8 **a)** 80 km
b) Point Fortin
c) 96 km

Chapter 6

Page 57

1 **a)** km^2 **b)** cm^2
c) cm^2 **d)** mm
e) cm^2 **f)** km^2
g) mm^2 **h)** m^2
i) mm^2 **j)** cm^2
2 **a)** 21 squares = 5.25 cm^2
b) 17 squares = 4.25 cm^2
c) 20 squares = 5 cm^2
d) 19.5 squares = 4.875 cm^2
e) 14 squares = 3.5 cm^2
f) 15 squares = 3.75 cm^2
g) 10 squares = 2.5 cm^2
h) 8 squares = 2 cm^2

Page 59

1 **a)** 27 cm^2 **b)** 24 cm^2
c) 21 cm^2 **d)** 36 m^2
e) 243 mm^2 **f)** 43.92 km^2
g) 36 cm^2 **h)** 144 mm^2
i) 9.61 m^2 **j)** 6.25 km^2
2 **a)** 20 cm^2 **b)** 36 cm^2
c) 13.5 m^2 **d)** 22.5 km^2
3 **a)** 19 cm^2 **b)** 36 cm^2
c) 85 m^2 **d)** 92 mm^2
e) 192 cm^2 **f)** 70 cm^2
4 3.8 cm
5 9 cm
6 120 cm^2
7 16.5 cm^2
8 39 cm^2

Looking back (page 62)

1 **a)** 42 mm^2 **b)** 210 mm^2 **c)** 66.5 cm^2
2 **a)** 16 cm^2 **b)** 65 mm^2 **c)** 46 m^2
3 864 cm^2
4 1 200 tiles
5 30.8 cm^2
6 50.41 cm^2
7

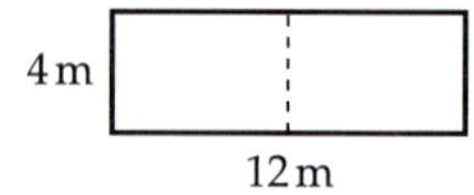

Chapter 7

Page 64

1 **a)** 3 **b)** −2 **c)** 6 **d)** −4
e) 36 **f)** −1 **g)** $\frac{1}{2}$ **h)** 0.5
2 **a)** 5 **b)** −12

Page 65

1 **Ages of some guests at the reunion party**

Guest	Age when Judith was born (years)	Age at time of party (years)	Age in two years' time (years)
Judith	0	10	12
Mr Mighty	80	90	92
James (Judith's brother)	Not yet born	8	10
Mr Mighty's wife	75	85	87
Judith's mother	26	36	38
Judith's father	28	38	40
Any guest	$p + 10$	p	$p + 2$

2 **a)** $x - 5$ **b)** $3x$ **c)** $x + 32$ **d)** $\frac{x}{7}$

Page 66

1 **a)** expression **b)** equation
c) expression **d)** equation
e) equation **f)** expression
g) equation **h)** equation
i) expression **j)** equation
k) equation **l)** expression

2 **a)** $20u - 18$ [u is the number thought of]
b) $v + 3v = 48$
[v is the number of marbles for Jojo's brother]
c) $\left(\frac{q}{2} + 2q\right)$ hours
[q is the number of hours that Mr Kings's son takes]
d) $t + (t + 1) + (t + 2) = 45$
[t is the lowest of the three consecutive numbers]
e) $g + 3(g + 2) = 50$
[g is the cost of one mango]
f) $b + (b - 5) + (b - 2)$
[b is Mark's age]

3 For example:
a) Five times the number minus 16 is equal to the number.
b) Nineteen years after the storm
c) Two times one boy's lunch money plus three times the other's is $150.00
d) Thirty items are left after 8 bags of cherries are sold and 5 have spoilt.
e) Finding angles in a right angle

Looking back (page 67)

1 **a)** 1 **b)** 3 **c)** 2 **d)** 2 **e)** 4
f) 1 **g)** 1 **h)** 2 **i)** 2 **j)** 3
k) 2 **l)** 3 **m)** 2 **n)** 1 **o)** 2

2 **a)** $b - 45$ **b)** $\frac{1}{2}b$ **c)** $b + 100$

3 **a)** $h + 0.6$ m **b)** $1.8s$ m
c) $\frac{x}{2}$ kg **d)** $(x + y + z)$ g

Chapter 8

Page 69

1 **a)** $p + p + p + p + p$
b) $k + k + k + k + k + k + k + k$
c) $b + b + b + b + b + b + b + b$
d) $g + g + g + g + g + g + g + g + g + g +$
$h \times h \times h \times h$
e) $y + y + y + y + y + y + y + y + y +$
$z + z + z + z$

2 **a)** $6a + 3p$ **b)** $4g + 5h$ **c)** m^7n^3

3 **a)** $11m$ **b)** $32p + 3q$
c) $25h + 10r + 6hr$ **d)** $17h^2 + 35p - p^2$
e) $3a^2 + ab + 10b^2$ **f)** $27k^3 + k^2$
g) $17efg - 16fg + ef$ **h)** $44p + 2pr + 4$

Page 70

1 **a)** 45 **b)** 133 **c)** 88
d) 120 **e)** −224

Page 72

1 **a)** $90 + 15k$ **b)** $3a + 3b + 3c$
c) $26r + 4p + 12 - 4q$ **d)** $\frac{3m}{2} - \frac{n}{2} + \frac{p}{2}$
e) $6g - 6h + 5j + 5k$

2 **a)** 5 **b)** $5a + b$ **c)** $4m$
d) $56 + 2a$ **e)** $11x + 3y + 5z$

3 **a)** 9 552 **b)** 1 218 **c)** 56 661
d) 935 868 **e)** 35 770
f) Students check their answers with a calculator.

Page 74

1 **a)** 39 **b)** 26 **c)** 38
d) 3 **e)** 25 **f)** 21

2 **a)** n **b)** $14m$ **c)** $3 - 8p$
d) $55 + 5m$ **e)** $3d + 2$

Looking back (page 74)

1 **a)** $3t - 3q$ **b)** $3r - p$ **c)** $x + 2y$
d) $b - 3t$ **e)** 0 **f)** $2y - 2u$

2 **a)** $y + y + y + y + y + y$
b) $z + z + z + y + y$
c) $a + a + a + a + a + a + a + b + b + b + c + c$
d) $s + s$
e) $4 \times 3 \times y \times y \times y$
f) $x + x \times x \times x + 4 \times 2$

3 **a)** $3 + 6y$ **b)** $28 + t - 2s$
c) $3w - 2v - 100$ **d)** $49t + 47$

4 **a)** 54 **b)** 74 **c)** 194 **d)** 94

5 **a)** $2 + 4c$ **b)** $8r - 28$
c) $84r - 12t$ **d)** $18a - 9$

6 **a)** 2 **b)** 1 **c)** 2 **d)** 3 **e)** 2
f) 3 **g)** 2 **h)** 1 **i)** 4 **j)** 1
k) 1 **l)** 4 **m)** 2 **n)** 2 **o)** 2
p) 4 **q)** 4

7 **a)** $14a$ **b)** $13m$ **c)** $54b$
d) $20a$ **e)** $17q$ **f)** $15k$

8 **a)** 6 **b)** 12 **c)** 12 **d)** 4 **e)** 6
f) 12 **g)** 9 **h)** 8 **i)** 12 **j)** 3
k) 2 **l)** 54 **m)** 8 **n)** 4 **o)** 7

9 **a)** $x + 6 = 12$ **b)** $3x + 4 = 22$
c) $x^2 - 7 = 29$ **d)** $2(x + 12) + 2 = 38$
e) $3x - 12 = 6$

Revise and consolidate 2

Page 78

1 **a)** 45 900 m **b)** 0.58 km
c) 920 cm **d)** 4.2 cm
e) 0.98 km **f)** 6730 mm
g) 5.647 m **h)** 0.0852 m
i) 60.9 cm **j)** 6.7 km

2 **a)** 92 cm **b)** 76 cm
c) 10.4 cm **d)** 17.8 cm
e) 4 000 cm **f)** 80 cm

3 32.3 m

4 **a)** 3 cm
b) 3 m
c) 350 g
d) 10 litres
e) Pupils' own answers.
f) Pupils' own answers.

5 **a)** T **b)** F **c)** F
d) T **e)** F **f)** T
g) T **h)** T

6 **a)** $n + 5$ **b)** $n + 14$
c) $6n + 7$ **d)** $8n + 12$

7 7

8 $6\frac{4}{7}$

9 $45t + 61s + 41p$

10 **a)** $10j + 8k$
b) $24k$
c) $10k + 11j$

11 **a)** T-shirt \$85 **b)** Skirt \$85
c) Shoes \$510 **d)** Laptop \$2010

12 1 836 cm^2

13 3.15 km^2

14 1.68 m^2

15 4 950 cm^2

16 48 000 m^2

Chapter 9

Page 82

1 Simple equations: d, e
Expressions: b, f
Formulae: a, c

2 **a)** pay $= 6c + r$ **b)** \$2 000 **c)** \$3 400

3 **a)** $6y = 24$ **b)** $y - 5 = 72$ **c)** $y + 13 = 29$
d) $2y + 8 = 22$ **e)** $y - 35 = 95$

4 **a)** $w - 4$ **b)** $2w - 4$ **c)** $2w - 4 = 30$

Investigation

a) \$10 260 **b)** \$10 170 **c)** \$9 840

Page 84

1 $\times 5$

2 **a)** $\div$ **b)** $\times$ **c)** $\div$
d) $+$ **e)** $\div$ **f)** $-$

3 **a)** 4 **b)** 1 **c)** 54
d) 13 **e)** 13

4 **a)** $n = m - 6$ **b)** $y = x + 10$

5 **a)** $\frac{x}{2} = 21$ **b)** $z = y + g$
c) $y = 2x$ **d)** $2x - 10 = 22$

Page 85

1 **a)** $q = 23$ **b)** $m = 19$
c) $p = 8$ **d)** $x = 8$
e) $x = 9$ **f)** $x = 7$
g) $r = 225$ **h)** $h = 2$
i) $y = 15$ **j)** $d = 50$

Page 87

1 **a)** $x = 7$ **b)** $y = 7$ **c)** $y = 5$
d) $m = 3.25$ **e)** $m = 22$

2 **a)** $y = 16$ **b)** $f = 18$ **c)** $a = 6$
d) $x = 4$ **e)** $n = 1$

3 **a)** $x = 6$ **b)** $x = 7$ **c)** $x = 8$
d) $x = -1$ **e)** $x = -2$ **f)** $x = 2$
g) $x = 2$ **h)** $b = 2$ **i)** $x = 5$
j) $f = 5$ **k)** $x = \frac{1}{2}$ **l)** $g = 4$
m) $h = 1$ **n)** $x = \frac{3}{5}$

4 **a)** $x = 4$ **b)** $a = 18$ **c)** $x = 0$
d) $d = 32$ **e)** $d = 40$ **f)** $y = 48$

5 **a)** $x = 4$ **b)** $x = 7$ **c)** $x = -1$
d) $p = 2$ **e)** $r = 6$ **f)** $y = 4$

6 3

7 12

8 8, 9, 10

9 15

10 Length = 33 cm, breadth = 11 cm

Looking back (page 88)

[1] 8	0			[2] 3	0	[3] 5
0		[4] 3	[5] 6	5		1
[6] 2	0	5	5		[7] 1	1
				[8] 1	2	
	[9] 3	3	3		[10] 3	[11] 1
[12] 1	4			[13] 1	4	0
[14] 7	0	0	0		[15] 5	1

Chapter 10

Page 91

1 Notes: \$100, \$20, \$10, \$5, \$1
Coins: 1c, 5c, 10c, 25c

2 a) A bank teller in Jamaica earns more dollars.
b) Selina in Barbados earns the least in dollars, but she may not be poorer than the others if her cost of living is lower.
c) No, because wages are also much higher.
d) 2% or $\frac{2}{100}$ for all of them.
e) The cost of living is the same in each country.

Investigation 1

Bahamas dollar
US dollar
Cuban peso
Euro
gourde
Netherlands Antilles guilder

Page 92

1 a) \$14 b) \$15
c) 85 kg d) 51 m
e) 60 kg f) 4.5 m
g) \$1080 h) \$240
i) 12 kg j) 12 kg

2 a) 7.7 km b) 33.6 m
c) 4.8 g d) \$86.70
e) 28.8 f) 4.5
g) 65 h) 7.2
i) 109.5
j) Students check their answers with a calculator.

3 a) 16 b) $87\frac{1}{2}\%$

4

	\$600	\$1 400	\$3 200
20%	\$120	\$280	\$640
35%	\$210	\$490	\$1 120
12%	\$72	\$168	\$384
51%	\$306	\$714	\$1 632
42%	\$252	\$588	\$1 344

5 a) 60% b) 65 c) 26

6 96 km

7 a) 14% b) 12 km

8 a) Mathematics 60%; Spanish 71%; English 80%; Science $53\frac{1}{3}\%$
b) English c) Brenda

9 71 cents

Page 95

1 a) 25% increase b) 50% increase
c) 12% decrease d) 1% increase
e) 3.5% decrease

2 a) \$44 b) \$450 c) \$60
d) \$150 e) \$14 300 f) \$68 000

3

Amount	Increase by 10%	Decrease by 20%
\$240	\$264	\$192
\$350	\$385	\$280
\$160	\$176	\$128
\$18	\$19.80	\$14.40
\$42	\$46.20	\$33.60

4 Photo frames \$20.00
Mugs \$9.60
Tea-towels \$20.80
Posters \$14.40
Fridge magnets \$2.72

5 25%

6 a) 10% b) 75% c) 17.5%

Page 97

1 a) \$96.45 b) \$132.16 c) \$79.00

2 a) \$192.55 b) \$207.95

3 a) \$955.00 b) \$1 098.25

4 a) \$24 b) 20%

5 a) \$63 b) \$37

6 a) \$33.75
b) It would have been \$3.75 cheaper

7 a) \$135.00 b) \$101.25 c) \$90.00
d) \$138.75 e) \$150.00

Page 99

1 \$105

2 \$2 934

3 a) \$5 800 b) \$2 800

4 a) \$2 305 b) \$177.08

5 a) \$6 000 b) \$1 500 c) $33\frac{1}{3}\%$

6 a) \$1 200 b) \$10 200 c) \$2 200

7 a) \$6 080 b) \$2 280 c) \$380

8 a) \$800 b) \$4 000 c) 100%

Page 101

1

Cost price	Selling price	Profit	Loss
\$300	\$320	\$20	
\$40	\$100	\$60	
\$25	\$20		\$5
\$1 350	\$1 275		\$75
\$545	\$480		\$65
\$798	\$840	\$42	
\$200	\$200	\$0	
\$2 500	\$3 100	\$600	
\$5 120	\$4 765		\$355
\$113	\$79		\$34

2 $20; $16\frac{2}{3}$%

3 a) loss b) $1 250

4 a) $187.50 b) $1 687.50

5 a) $7 000 b) 7.29%

6 $6 000

7 $3 000

8 a) Harry $600; Paul $750
b) Paul's
c) Paul

Page 102

1 a) $360 b) $72
c) $250 d) $1 536
e) $8 100 f) $3 900
g) $64 h) $7 020
i) $204 000 j) $750

2 a) $1 560 b) $672
c) $1 250 d) $4 736
e) $26 100 f) $23 900
g) $464 h) $16 020
i) $324 000 j) $5 750

Looking back (page 104)

1 36

2 a) 45% b) 260

3 50

4 a) $1 000 b) $\frac{8}{25}$ c) $450

5 a) $296
b) $340.40
c) $9.60
d) 3 shower curtains $123.68; 120 sweets $6.80; 4 tins of air freshener $51.00; coffee maker $63.75; 3 packets of flour $6.38

6 a) $1 b) $2.25 c) $5.25

7 a) $3 780 b) $1 380

8 $2 100; 28%

9 $2 000

10 a) $760 b) $9 600

Revise and consolidate 3

Page 106

1 a) $x = 9$ b) $h = 56$
c) $t = 20$ d) $m = 31$
e) $x = 9$
f) Students check their solutions.

2 a) 7 times a number, add 5, result is 9
b) 4 less than a number is 12
c) 8 times a number, divided by 6, result is 2
d) a number taken from 25, result is 4
e) 9 less than a number = 4 times the number

3 a) $5x = 20$ b) $x = 4$ kg

4

	Cost price	Selling price	Profit or loss in $	% Profit or loss
a)	$100	$120	$20 P	20% P
b)	$1.00	$1.50	$0.50 P	50% P
c)	$55	$66	$11 P	20% P
d)	$160	$150	$10 L	6.25% L
e)	$50	$56	$6 P	12% P
f)	$1.60	$1.40	20¢ L	12.5% L
g)	$10	$11.25	$1.25 P	12.5% P
h)	$0.50	$0.30	20¢ L	40% L
i)	$10	$13	$3 P	30% P
j)	$1.67	$1.60	7¢ L	4% L

5 a) $2 280 b) $190

Chapter 11

Page 110

1 a) ray b) line
c) line segment d) ray
e) ray f) line
g) line segment h) line segment

2 a) ray b) line segment c) ray

3 a) 3 cm b) 7 cm c) 5.5 cm
d) 10 cm e) 4.5 cm

4 Pupils' own accurate drawings.

5 Pupils' own accurate drawings.

Page 111

1 ∠EHG or ∠GHE ∠MON or ∠NOM
∠RST or ∠TSR ∠PXZ or ∠ZXP
∠VYP or ∠PYV ∠QRS or ∠SRQ

2 ∠BAC, ∠BEC, ∠CEF, ∠JDE, ∠DEC, ∠EFH, ∠FHG, ∠EKM, ∠EKL, ∠DEK, ∠LKM, ∠ABE, ∠DEB, ∠BEK, ∠ACE, ∠FEB, ∠CEK, ∠FEK

3 Pupils' own answers.

Page 113

1 Pupils' own accurate drawings.

2 Pupils' own work.

3 a)

Acute	Obtuse	Reflex
18°, 70°, 36°, 86°, 5°	115°, 160°, 93°, 155°, 143°	215°, 185°, 340°, 295°

b) Pupils' own accurate drawings.

4 a) Pupils' reasonable estimates of angles.
b) $a = 47°$ $b = 123°$ $c = 82°$
$d = 286°$ $e = 248°$ $f = 153°$
$g = 102°$ $h = 84°$ $i = 79°$
$j = 230°$ $k = 53°$ $l = 120°$

Page 115

1 **a)** opposite
b) supplementary
c) complementary
d) x – opposite, a/b – supplementary
e) x/y and a/b – supplementary
f) complementary

2 **a)** e.g. ∠EGC and ∠KGH
b) e.g. ∠BEG and ∠JEG
c) e.g. ∠EGC and ∠CGH

3 **a)** $a = 110°, b = 70°$ **b)** $p = 115°$
c) $x = 55°$ **d)** $c = 36°$
e) $e = 80°$

4 **a)** $a = 40°$ **b)** $b = 72°$
c) $c = 51°, d = 270°$ **d)** $e = 132°$
e) $f = 54°$ **f)** $a = 100°$
g) $b = 40°, c = 140°, d = 40°$
h) $e = 135°, f = 45°, g = 135°$

5 $p = 56°, q = 34°$

6 $x = 40°, y = 50°, z = 40°$

7 **a)**

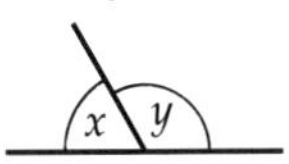

b)

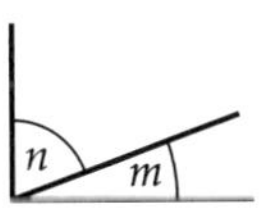

c)

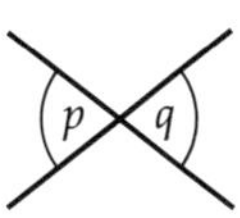

Looking back (page 117)

1 **a)** Any three of: A, B, C, D, E, F
b) Six angles: ∠CAD, ∠DAE, ∠EAB, ∠CAE, ∠DAB, ∠CAB
c) Any of: AE, EF, CF, AC
d) AB or AD

2 Pupils' own drawings.
a) obtuse **b)** obtuse
c) reflex **d)** straight
e) acute **f)** right angle

3 **a)** 71° **b)** 125° **c)** 28° **d)** 90°

4 **a)** $p = 16°$ (supplementary)
$q = 164°$ (supplementary)
$r = 16°$ (supplementary)
b) $v = 42°$ (supplementary)
$u = 94°$ (supplementary)
$t = 86°$ (opposite/supplementary)
c) $x = 32°$ (complementary)
d) $x = 90°$ (supplementary)
$y = 65°$ (supplementary)
e) $a = 100°$ (supplementary)
$b = 100°$ (supplementary)
$c = 80°$ (supplementary)
f) $x = 90°$ (supplementary)
$y = 45°$ (supplementary)

Chapter 12

Page 118

Students' own answers

Page 119

1

Solid	Faces	Edges	Vertices
Cone	2	1	0
Cuboid	6	12	8
Cube	6	12	8
Cylinder	3	2	0
Square-based pyramid	5	8	5
Hemisphere	2	1	0
Sphere	1	0	0

Page 120

Students' own answers

Page 121

1 A to 6
B to 3
C to 1
D to 5
E to 2
F to 4

2 Students' own answers

3 Students' own answers

Looking back (page 122)

1 In order: sphere, cylinder, cone, cuboid, prism, pyramid, cube, hemisphere

2 **a)**

	A	B	C	D	E
Faces	9	4	2	6	3
Edges	16	2	1	12	2
Vertices	9	none	none	8	1

b)

	A	B	C	D	E
Shapes used	cuboid and pyramid	part of a cone	cylinder and hemisphere	prism	cylinder and cone

c)

	A	B	C	D	E
Curved faces	none	2	1	none	2

3 a)

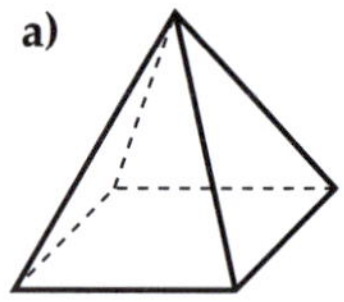

b)

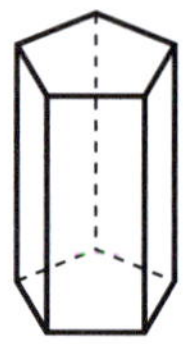

c)

4

Solid	Number of straws	Number of beads
A	12	8
B	12	7
C	12	8
D	9	6
E	12	8

5

Parallel	Perpendicular	Neither
A	B	H
C	E	J
D	F	K
I	G	
L		

Chapter 13

Page 125

1 **a)** right-angled isosceles
b) obtuse-angled scalene
c) equilateral
d) equilateral
e) acute-angled isosceles
f) acute-angled isosceles
g) equilateral
h) obtuse-angled scalene

2 Pupils' own correct drawings.

3 A Equilateral
B Scalene acute
C Right-angled
D Scalene obtuse
E Scalene obtuse
F Right-angled isosceles
G Obtuse isosceles

4

Name of triangle	Size of each angle	Sum of the angles
Right-angled	90°, ∗, ∗	180°
Right-angled isosceles	90°, 45°, 45°	180°
Obtuse scalene	∗, ∗, ∗	180°
Acute scalene	∗, ∗, ∗	180°
Acute isosceles	∗, ∗, ∗	180°
Obtuse isosceles	∗, ∗, ∗	180°
Equilateral	60°, 60°, 60°	180°

∗ = Pupils' own answers

Page 128

1 **a)** $a = 72°$ **b)** $d = 22°$
c) $f = 76°$ **d)** $g = 125°$
e) $h = 189°$ **f)** $i = 47°$
g) $j = 75°$ **h)** $k = 96°$

2 **a)** $a = 78°, b = 23°, c = 79°$
b) $d = 115°, e = 25°, f = 25°$

3 **a)** $x = 35°$ **b)** $x = 65°$ **c)** $x = 57°$

Page 129

1 **a)** rectangle
b) parallelogram
c) kite
d) square
e) rhombus

2

Quadrilateral	Opposite sides are always parallel and equal (Yes/No)	Opposite angles are always equal (Yes/No)	Angles are always right angles (Yes/No)
Parallelogram	Yes	Yes	No
Rectangle	Yes	Yes	Yes
Square	Yes	Yes	Yes
Rhombus	Yes	Yes	No
Kite	No	No	No
Trapezium	No	No	No

3 **a)** square **b)** trapezium
c) rhombus **d)** kite

4 **a)** square or rhombus
b) square, rhombus, rectangle, parallelogram, kite
c) square, rhombus, rectangle, parallelogram
d) square, parallelogram, rectangle, rhombus

5 Pupils' own drawings.

6 **a)** true
b) false
c) false
d) true
e) true

7 Pupils' own drawings.
  a) either square or isosceles triangle
  b) either rectangle or isosceles triangle
  c) either parallelogram or rhombus
8 Pupils' own accurate drawings.

Page 133

1 **a)** $a = 120°, b = 120°$
  b) $x = 63°$
  c) $x = 32°$
2 **a)** $x = 25°, y = 58°$
  b) $a = 135°, b = 45°, c = 45°$
  c) $x = 58°$
3 **a)** $x = 50°$
  b) $x = 36°$

Page 135

1
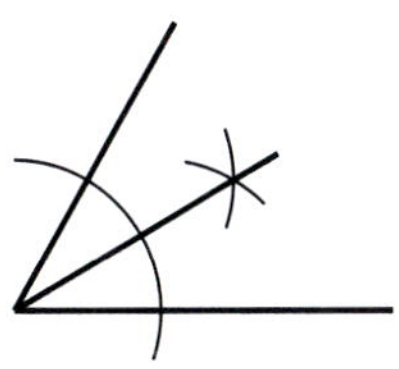

2
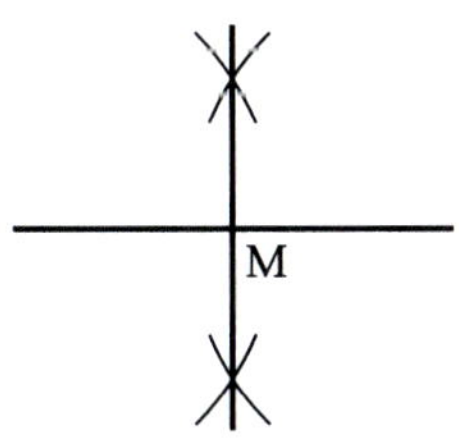

3
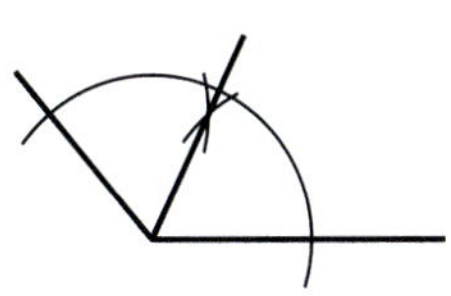

4
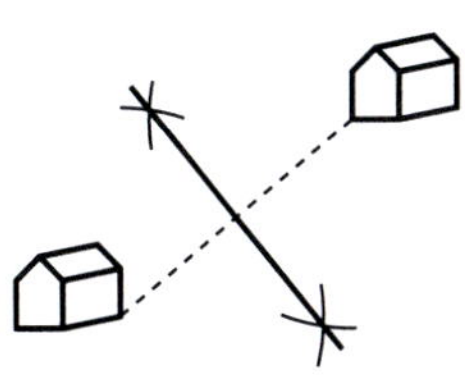

5
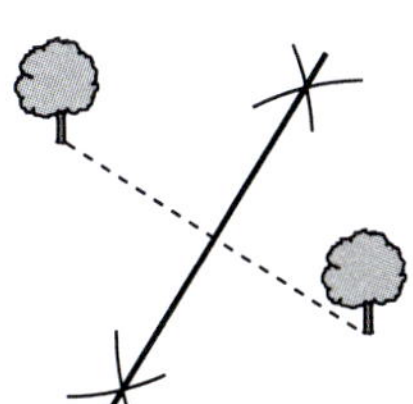

Looking back (page 136)

1 **a)** Scalene
  b) Isosceles
  c) Equilateral
2 **a)** $x = 40°$ **b)** $x = 59°$
  c) $x = 64°$ **d)** $x = 82°$
  e) $x = 21°$ **f)** $x = 130°$
3 **a)** Trapezium
  b) Rhombus
  c) Parallelogram
  d) Kite
  e) Rectangle
  f) Square
4 Students' own drawings.
5 **a)** $d = 75°$
  b) $d = 73°$
  c) $d = 129°$
  d) $d = 115°$
  e) $x = 55°, y = 70°$
  f) $x = 80°, y = 80°$
  g) $x = 50°, y = 40°$
  h) $w = 155°, x = 35°, y = 50°, z = 25°$
  i) $x = 70°, y = 110°, z = 70°$
6
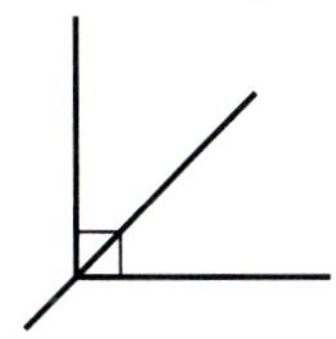

7
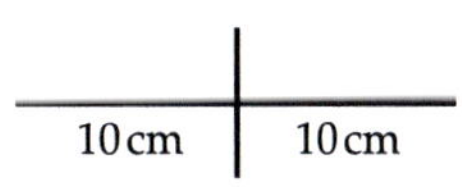

8
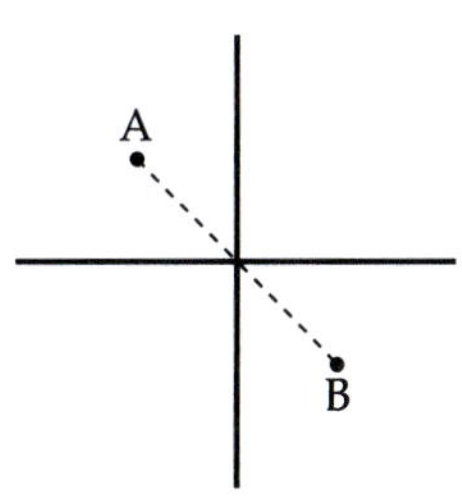

Revise and consolidate 4

Page 139

1 A cuboid
  B tetrahedron
  C cylinder
  D cone
  E square-based pyramid
  F triangular prism
  G pentagonal prism
  H hexagonal prism
  I hexagonal-based pyramid
  J cube
2 **a)** B **b)** H **c)** F **d)** I **e)** G
3 ∠RQS or ∠SQR acute
  ∠ABC or ∠CBA acute
  ∠DEF or ∠FED obtuse
  ∠HIJ or ∠JIH obtuse
  ∠KLM or ∠MLK reflex
  ∠NOP or ∠PON reflex

4 Students' own drawings.

5 **a)** Scalene **b)** Isosceles
c) Scalene **d)** Right-angled

6 **a)** Trapezium **b)** Rhombus
c) Parallelogram **d)** Kite

7 Students' own drawings.

8

Angle	Complement	Supplement
50°	40°	130°
90°	0°	90°
60°	30°	120°
40°	50°	140°
88°	2°	92°
89°	1°	91°
45°	45°	135°

9 **a)** The angles should sum to 360° as they are a complete rotation but they sum to 350° so it is mathematically impossible.
b) The angles on a straight line should sum to 180° but the angles on one of these straight lines sum to 170° so it is mathematically impossible.

10 **a)** $\angle AGC = 90°$ **b)** $\angle EGD = 140°$
c) $\angle ADC = 138°$

11 **a)** 45° **b)** 70°
c) 50° **d)** 25°
e) 60° **f)** 45°
g) 43° **h)** 60°

12 $p = 56°$ (complementary to 34°)
$q = 34°$ (sum of angles in a triangle = 180°, $180° - 90° - 56° = 34°$)

Chapter 14

Page 143

1 Students' own answers

2 **a)**

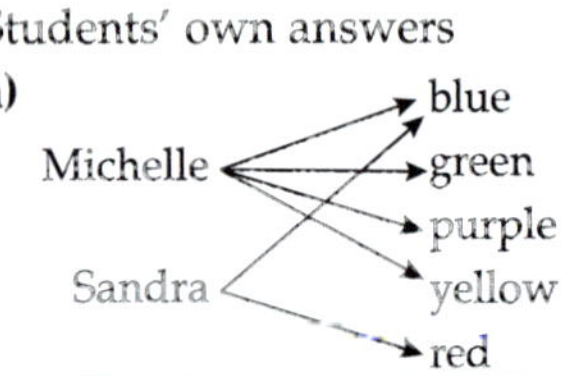

b)

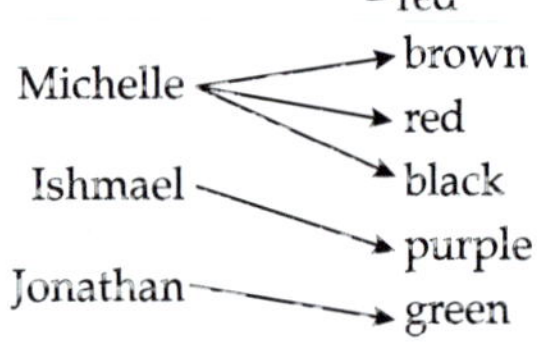

3 **a)**

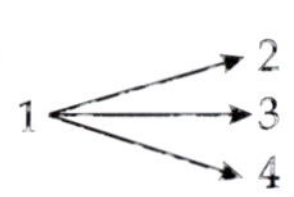

Domain: 1
Range: 2, 3, 4

b)

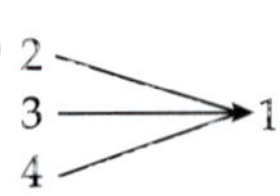

Domain: 2, 3, 4
Range: 1

c)

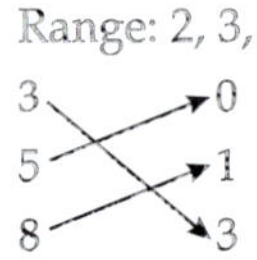

Domain: 3, 5, 8
Range: 0, 1, 3

d)

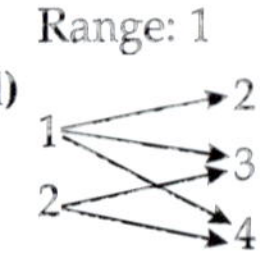

Domain: 1, 2
Range: 2, 3, 4

Page 143

1 **a)** Domain: 3, 4, 6; range: 7, 9, 13
b) Domain: 2, 3, 4; range: 8, 27, 64
c) Domain: 1, 2, 6; range: $\frac{1}{2}$, 2, 18

2 (2, 4), (5, 10), (10, 20), (12, 24), (200, 400)

3 **a)** Is half of
b) Is 1 less than
c) Is the square of
d) Is 18
e) Is one tenth of

Page 145

1 **a)** many-to-one **b)** many-to-one
c) many-to-many **d)** one-to-many
e) one-to-one **f)** many-to-many

2 **a)** many-to-many

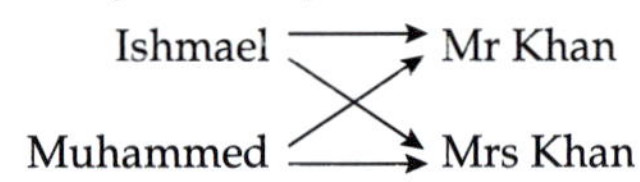

b) one-to-many

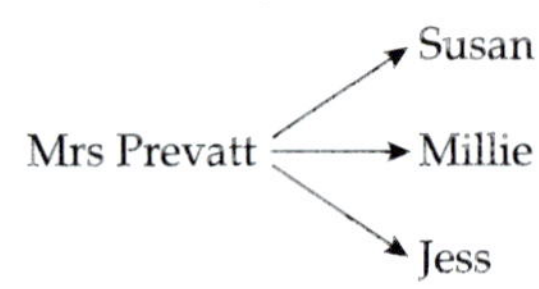

c) many-to-many

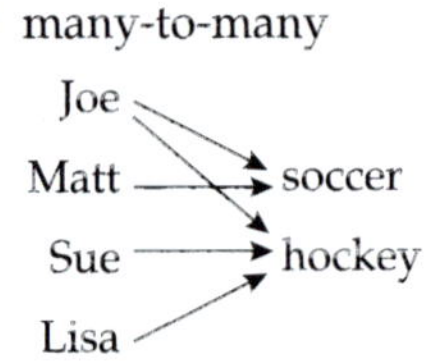

d) one-to-one

e) many-to-one

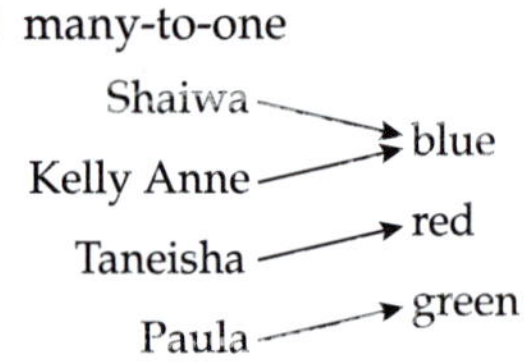

f) many-to-many

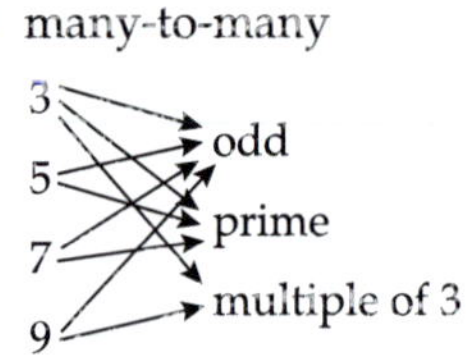

3 **a)** one-to-one **b)** one-to-many
c) many-to-one **d)** many-to-many

Page 146

1 **a)**

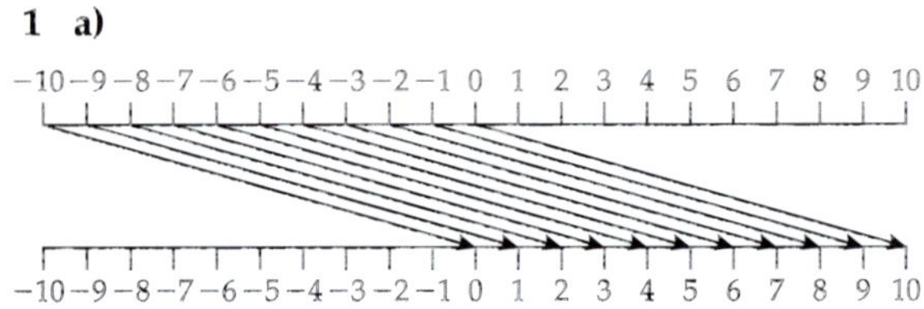

b)

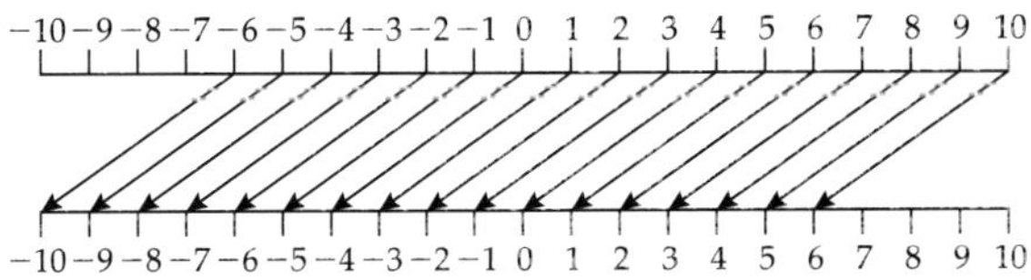

c)

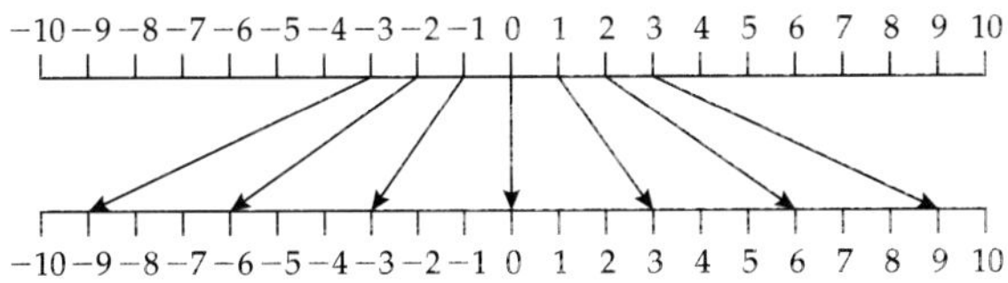

d)

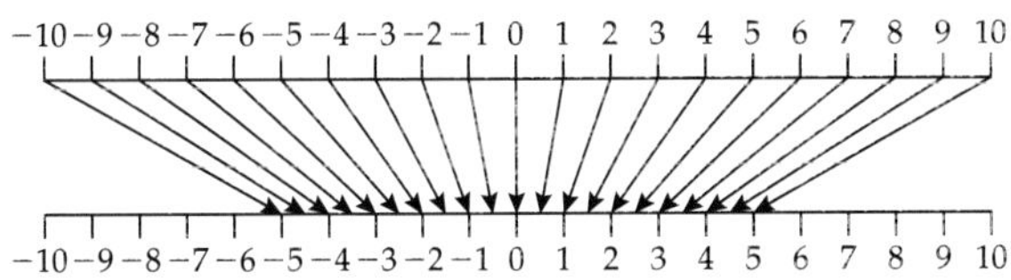

2 **a)** $1 \rightarrow 1$
$2 \rightarrow 6$
$3 \rightarrow 11$
$4 \rightarrow 16$

b) $0 \rightarrow 7$
$5 \rightarrow 8$
$10 \rightarrow 9$
$15 \rightarrow 10$

c) $12 \rightarrow 1$
$23 \rightarrow 2$
$34 \rightarrow 3$
$45 \rightarrow 4$

d) $0 \rightarrow 0$
$5 \rightarrow 2$
$10 \rightarrow 4$
$15 \rightarrow 6$

3 **a)** $x \rightarrow x + 3$ **b)** $x \rightarrow x - 2$

4 **a)** $x \rightarrow x + 6$ **b)** $x \rightarrow x - 1$

c) $x \rightarrow 5x$ **d)** $x \rightarrow \frac{x}{6}$

5 **a)** **i)** larger **ii)** smaller **iii)** smaller **iv)** larger

b) Students' own answers

6 **a)** **i)**

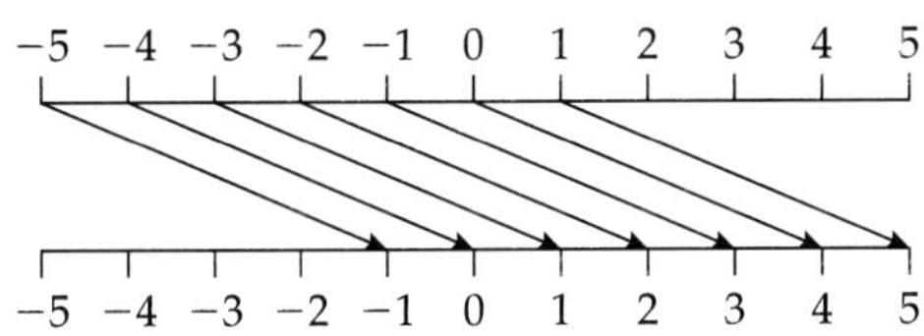

ii)

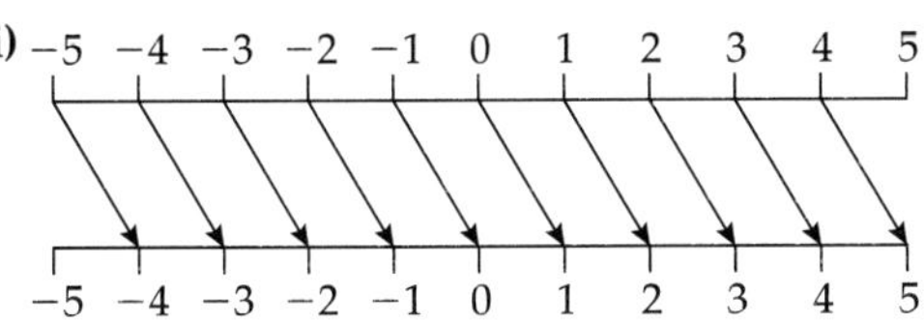

iii)

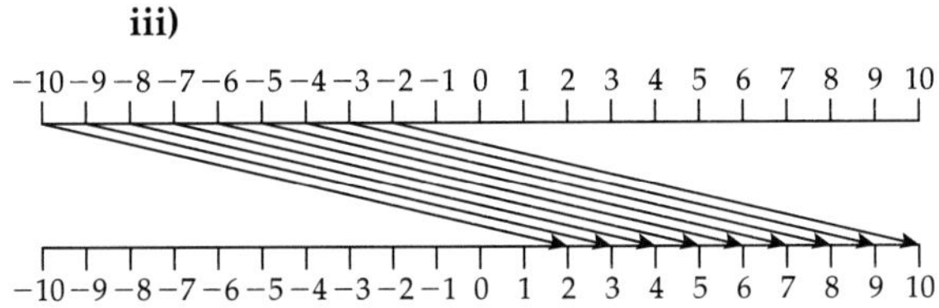

b) They are parallel.

c) The same number is added each time.

7 Students' own answers

8 **a)** $x \rightarrow 7x$ **b)** 35 miles

c) 280 miles **d)** 6 litres

Page 149

1 **a)** (1, 6) **b)** (4, 6) **c)** (6, 7)
d) (3, 5) **e)** (6, 5) **f)** (1, 4)
g) (3, 3) **h)** (5, 3) **i)** (1, 2)
j) (3, 2) **k)** (6, 2) **l)** (7, 2)
m) (0, 1) **n)** (2, 1) **o)** (6, 0)

2 **a)** N **b)** A **c)** I
d) G **e)** K **f)** B
g) E **h)** H **i)** J
j) C **k)** D **l)** L
m) O **n)** F **o)** M

3 Jack's village = (A, 1)
Beach Hut = (C, 1)
Pebbles Hill = (B, 2)
Walker Hall = (B, 3)
Youth Halt = (B, 3)

Page 150

1 **a)**

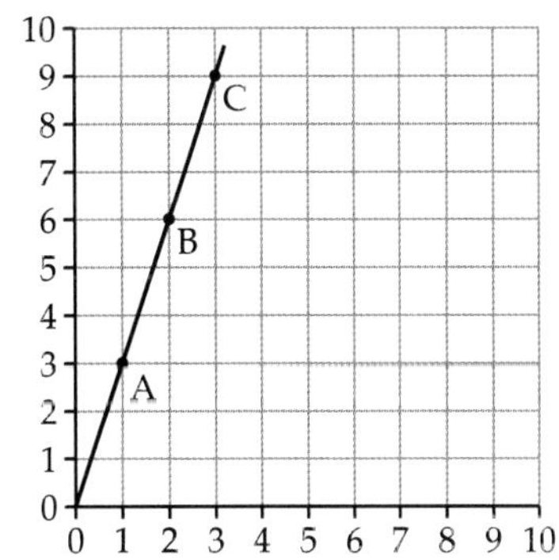

b) e.g. (4, 12) (5, 15) (25, 75)

c) $y = 3x$

2 **b)** $x = 5$

3 **b)** square

4 **a)**

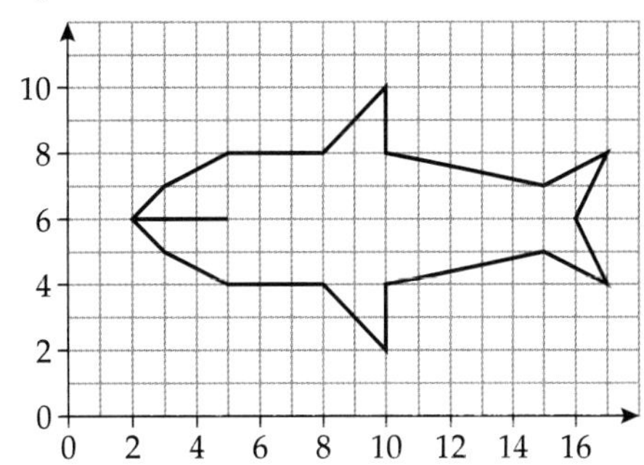

b)

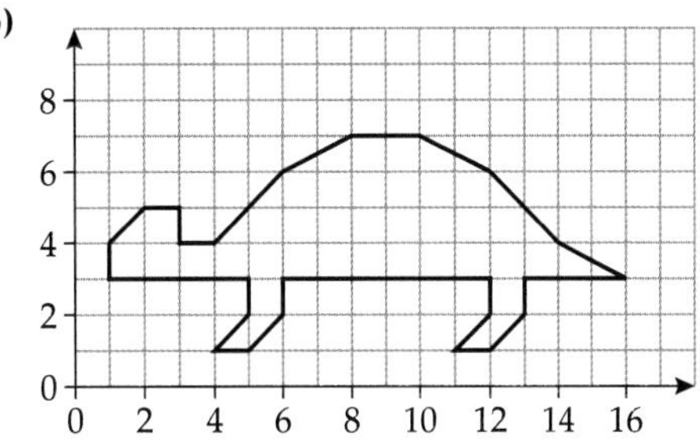

Looking back (page 151)

1 Pupils' own drawings.

2 Hexagon

3 c) On the x-axis $y = 0$
On the y-axis $x = 0$
The coordinates of the origin are (0, 0).

4 a) (1, 6) (3, 6)
b) —
c) (1, 5), (3, 7)
d) (1, 3), (3, 7)
e) (1, 2), (3, 8)

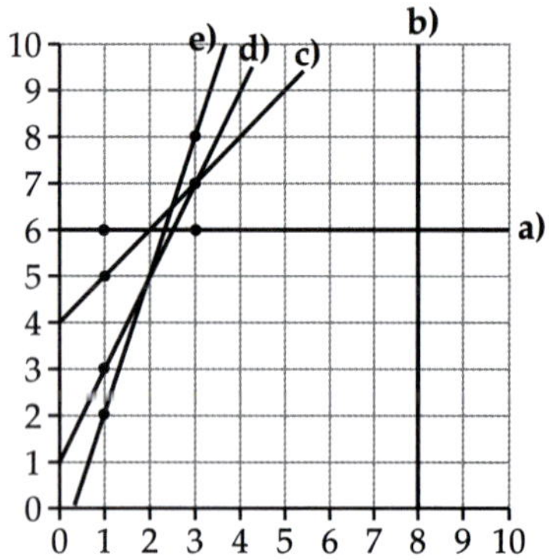

5

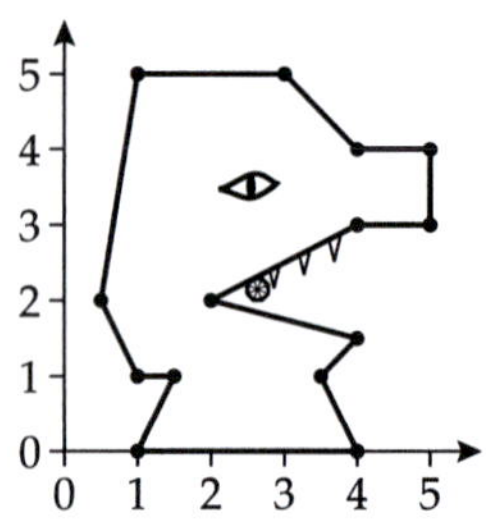

Chapter 15

Page 154

1 a) survey
b) experiment
c) survey
d) secondary source
e) survey
f) experiment

2 The experiment did not include enough throws of the dice. A fair conclusion cannot be made with ten throws.

3 a) Most of the people outside the cricket ground would more than likely enjoy cricket.
b) Need to get data from people who like to read as well as those who don't. People in a library would all like to read.
c) If in the queue then the response could only be yes.

4 a) metres and centimetres
b) kilometres
c) hours
d) years and kilograms
e) centimetres
f) hours and years

Page 156

1 a) 60 b) 110

2 a) 420 b)

3 a)

Fruit	Frequency
banana	6
apple	7
grapes	3
cherry	4

b) apple
c) grapes

4

Allison	■ ■ ■
Ronald	■
Grace	■ ■ ■ ■ ■
Greg	■ ■ ■ ■
Barry	■ ■

Page 158

1

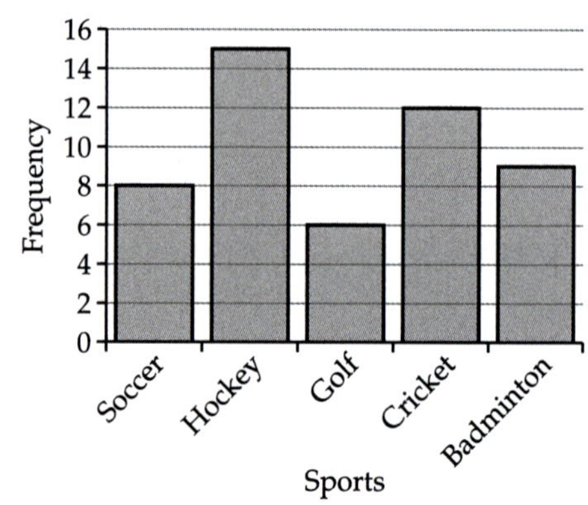

2 a)

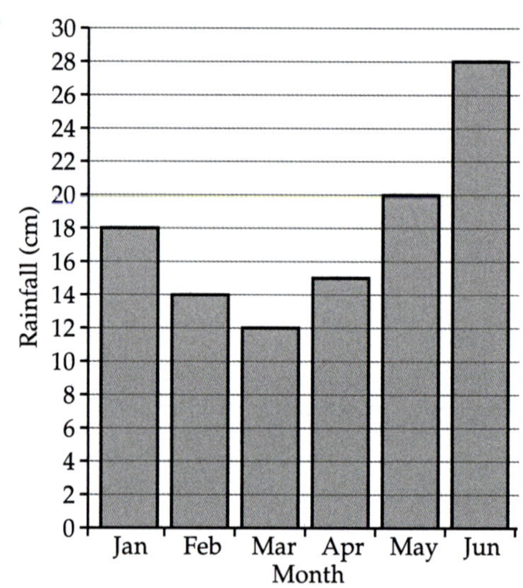

b) June
c) January, May and June

3 a) Yellow b) 20
c) Yellow and Blue; Red and Purple

Page 159

1 **a)** chocolate
b) coconut
c) 120

2
Barbados	96°
Trinidad and Tobago	72°
Grenada	24°
Guyana	48°
Jamaica	120°

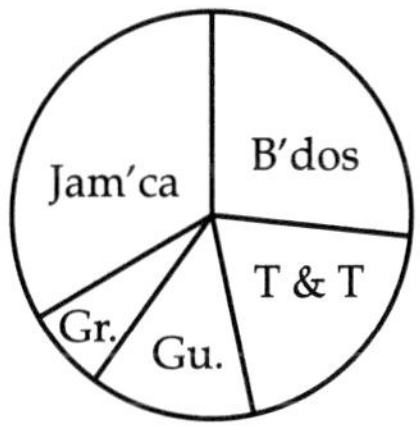

3 **a)** 90 children
b) 4°
c) Students' own drawings

4 **a)** 6 **b)** 3 **c)** 4
d) Students' own answers

Looking back (page 162)

1 **a)** pie chart **b)** pictograph
c) bar chart **d)** bar chart

2 **a)**

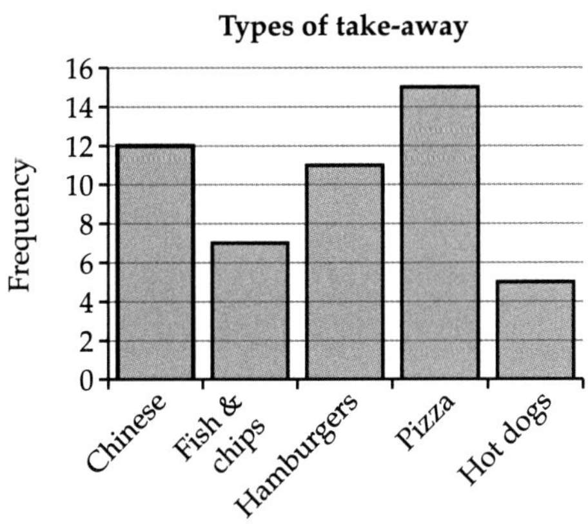

b) $\frac{15}{50}$

3

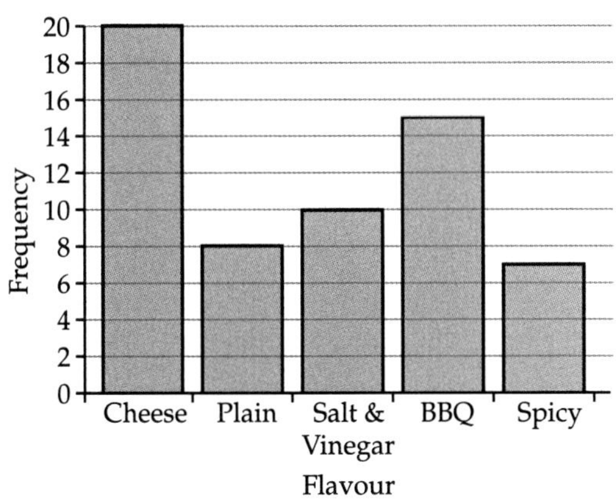

4 **a)** 30 **b)** 24 **c)** 9

5 **a)** Suzuki **b)** 8 **c)** 64
d)

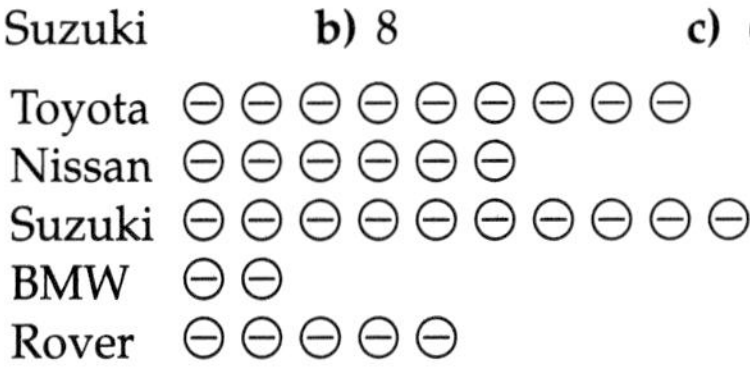

Chapter 16

Page 164

1 **a)** 6 **b)** 0 **c)** 3 **d)** 2.4 **e)** 4.6

2 white

Page 165

1 **a)** median 3; range 4 **b)** median 9; range 5
c) median 12.5; range 5 **d)** median 1.2; range 1.1

2 **a)** median mark 7 **b)** range 5

3 median 57.5

Page 166

1 **a)** 16 **b)** 8 **c)** 11

2 $18.00

3 $x = 13$

4 70%

5 20.04 s

Page 168

(Some questions have many different answers. These are some possibilities).

1 **a), c)** and **g)** are all primary sources, either by performing an experiment or using a questionnaire.
b), d), e) and **f)** are all secondary sources.

2 **a)** How many times do you play sport each week?
0 1 2 3 4 5+
What sports do you play?
Football Swimming Tennis Other
Do you think children play enough sport?
Yes No Don't know
b) He should stand in different places and at different times. He should ask a lot more pupils and more variety of ages.
c) He might want to talk again to pupils that he has already asked, or split his data into different age groups or into boys and girls.

3 Crime statistics

4 You would use scatter graphs to plot the 3 different pairs of variables. You should have included how many people you would ask. Results can be recorded to the nearest second.

Time taken (secs)	0–10	10–20	20–30	30–40	40–50	50+

5 Lata needs to find out which teams have changed division, perhaps between this season and last season. She needs to find out information about the size of the crowd at different matches for these teams, perhaps using an Internet sports site. Once she has collected her data she can calculate statistics to compare the size of crowd between the two divisions for each team.

Page 170

Students' own plans

Page 172

1 Time spent on an investigation

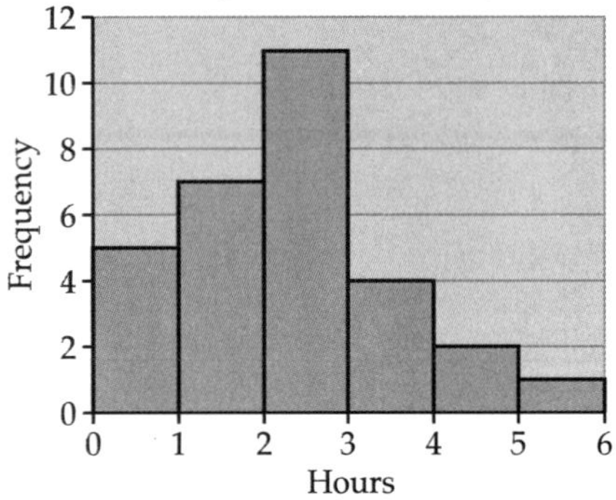

2 Heights of boys in Form 1

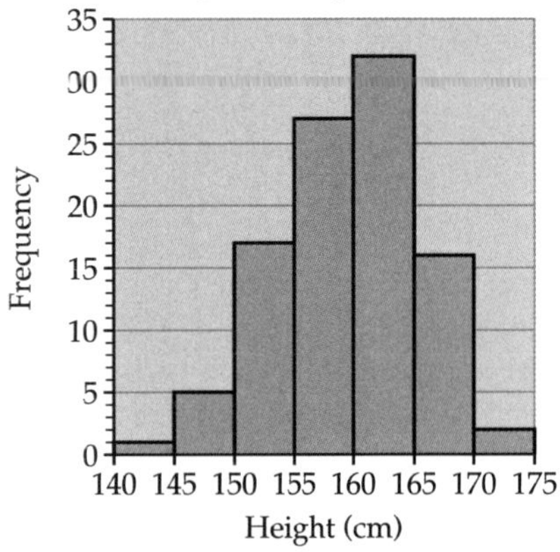

3 Type of part-time job done by pupils

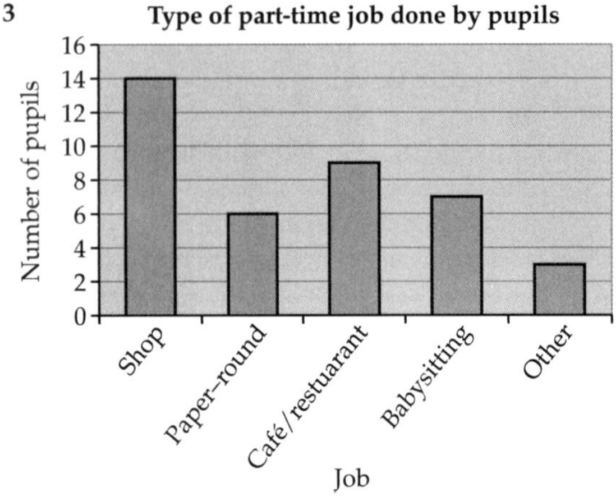

Hours worked by pupils

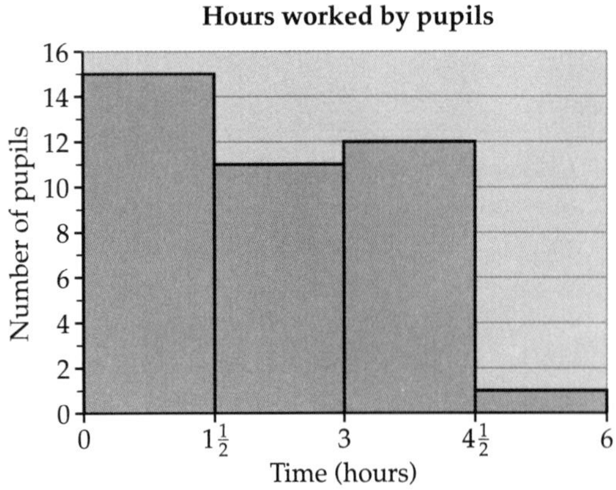

Amount per hour earned by pupils

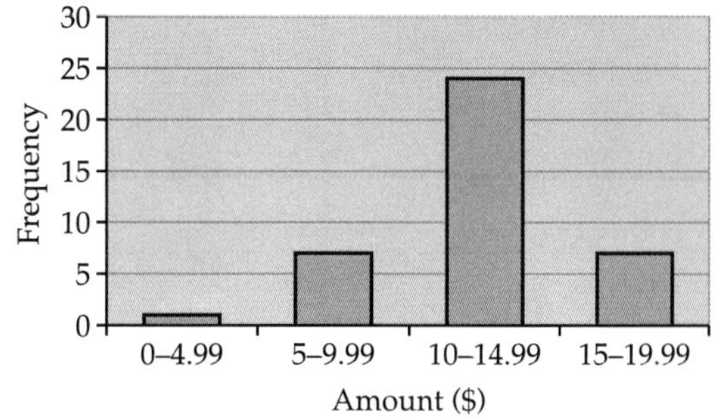

4

Time (hours)	8 – 10	10 – 12	12 – 14	14 – 16	16 – 18	18 – 20
Frequency	2	5	7	2	3	1

Time taken for long-distance swim

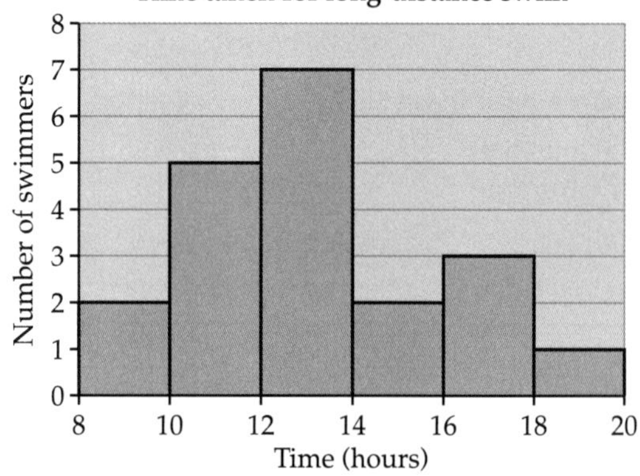

Page 174

1 a) 1 – 11 b) 8%

c) The remainder will be people who spend 0 minutes getting to work (people who work from home).

d) A greater percentage of women spend 30 minutes or less travelling to work, but a greater percentage of men spend 0 minutes travelling to work.

2 a) 25 – 30 seconds, 30 – 35 seconds, 35 – 40 seconds, 40 – 45 seconds

b) 25 c) 35 – 40 seconds

d) 12 e) 20

f) As the data is grouped it is not possible to see if any pupil had a time of exactly 25 or 45 seconds.

g) The data is grouped.

3 Students' own answers

Looking back (page 175)

1 a) mean: 4 median 3.5

mode 6 range 4

b) The store should stock the most popular sizes, i.e. the mode and other sizes which were bought regularly.

2 a)

Number in box	Frequency
44	1
45	3
46	3
47	1
48	3
49	2
50	4
51	2
52	1

b) mean: 48 median: 48 mode: 50
c) The mode was used to describe the average.
d) Students' own answers

3 **a)** mode **b)** mode **c)** mode

4 **a)** mean **b)** mode **b)** mean
d) mode **e)** mode

5 **a)** mode: 32
mean: 53.2
median: 67
b) the median

6 **a)** 20 **b)** 4 **c)** 129 **d)** 6.45
e) Above average, because his average as a single player is nearly as high as the mean for teams.

Test your knowledge

Test 1

1 d
2 b
3 c
4 c
5 d
6 **a)** 512 813 **b)** 513 000
7 **a)** one of 16, 18, 20, 22 **b)** one of 63, 77, 91
c) one of 16, 20, 40 **d)** 53, 59
8 **a)** **i)** 3 **ii)** 36 **b)** **i)** 8 **ii)** 80 **c)** **i)** 12 **ii)** 420

Test 2

1 b
2 c
3 d
4 **a)** 8 246 **b)** 2 149 **c)** 15 808 **d)** 24
5 **a)** 13 km **b)** 7 km
6 **a)** \$2.50 **b)** \$60.00 **c)** \$198.00
7 **a)** 27.5 **b)** 14.72 **c)** 48
8 d
9 **a)** \$25.05 **b)** \$53.75
c) \$103.65 **d)** \$104.00
10 a

Test 3

1 d
2 a
3 c
4 \$14 540
5 \$5 550
6 **a)** \$23.75 **b)** \$256.25
7 b
8 b
9 c
10 a
11 a
12 d

Test 4

1 **a)** $y + 4$ **b)** $x - 15$ **c)** $3a$ **d)** $d - 5$ **e)** $p - q$
2 **a)** 10 **b)** 13
3 \$580.75
4 b
5 c
6 b
7 a
8 1.6 m^2
9 10.312 kg

Test 5

1 b
2 d
3 b
4 c
5 d
6 a
7 80 000 km
8 a
9 \$51.10
10 \$560

Test 6

1 c
2 d
3 b
4 Pupils' own answers.
5 **a)** 90° **b)** 90° **c)** 330°
6 **a)**

b)

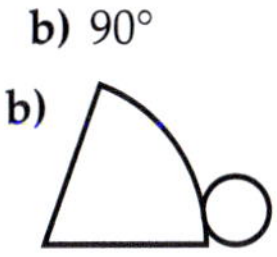

c)

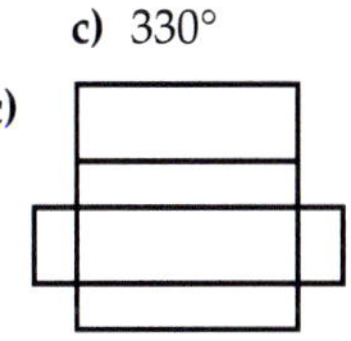

7 **a)** 6 **b)** 5 **c)** 4

Test 7

1 b
2 d
3 a
4 a
5 a
6 a
7 b
8 c

9 d

10 a) $6b + 6c$ b) 30

11 $\$20d$, $\$20d - 30$, \$9

12 a) 30° b) 25° c) 40°

Test 8

1 b

2 b

3 a

4 b

5 b

6 d

7 a) \$23.75 b) $4\frac{1}{2}$ hours

8 6, 21, 33, 63 → 4, 9, 31, 61

9 a) Students' own drawings

b) isosceles c) 2

10 (4, 7)

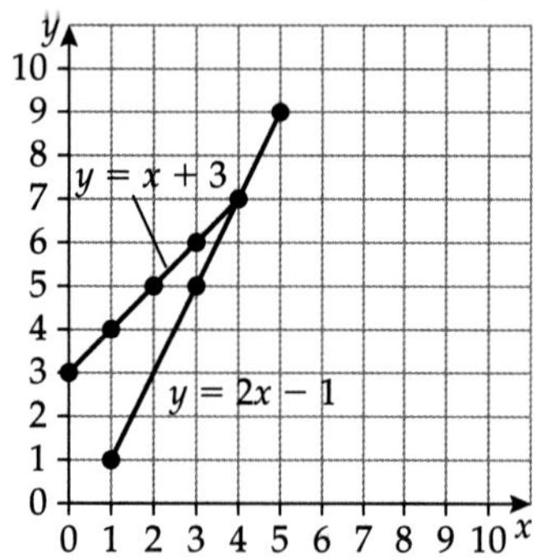

Test 9

1 d

2 a

3 Maths 88°
English 60°
Science 72°
History 56°
Geography 84°

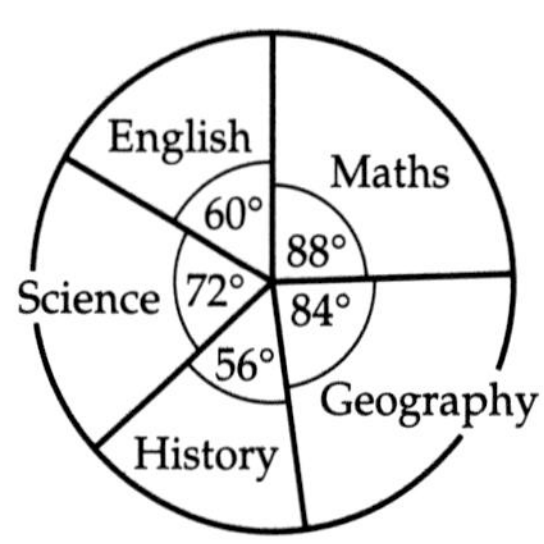

4 a)

Age	Tally	Frequency
13	𝍸	5
14	𝍸 𝍸 𝍸	15
15	𝍸 𝍸	10
16	𝍸 I	6

b)

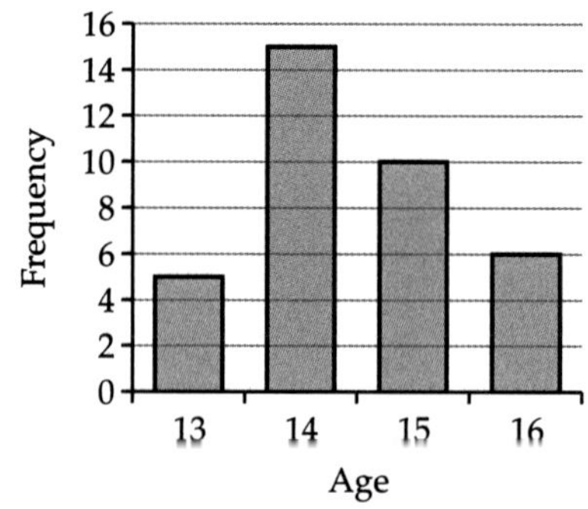

5 40.25 cm

6 177 cm

Test 10

1 b

2 d

3 d

4 c

5 b

6 d

7 c

8 c

9

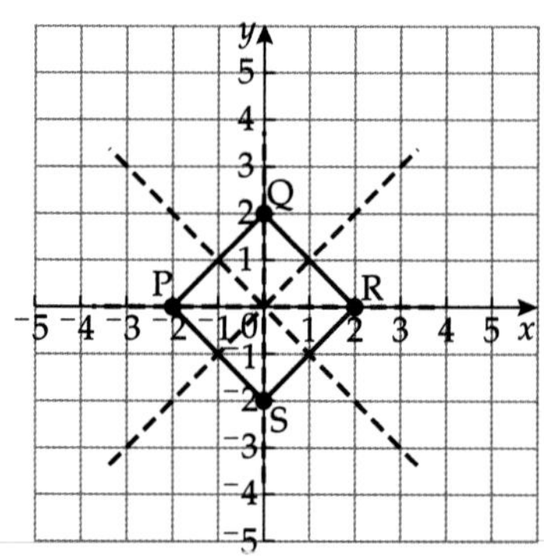

10 6 cm

Index